Iron's Parameter: Shaping Oceans

Iron's Parameter: Shaping Oceans

Navya

Copyright © 2024 by Navya

All rights reserved. No part of this book may be reproduced in any manner
whatsoever without written permission except in the case of brief quotations embodied in critical articles and reviews.
First Printing, 2024

Horace, old friend, life has taken us on quite
a journey, full of twists and turns

Table of Contents

1. Introduction

1.1 Background

Iron is a crucial limiting micronutrient for phytoplankton, and it has been linked to processes such as photosynthesis, respiration and nitrogen fixation (Morel & Price 2003). Iron was showcased to be an important limiting nutrient for productivity in the 1980's; however, it was only in the early 2000's that iron was included as a variable in global ocean models (Tagliabue *et al.* 2017). In describing the dynamics of iron, biogeochemical models employ a host of partial differential equations (PDEs) to represent various chemical, physical and biological processes. However, there is no generally accepted set of equations to model a marine ecosystem and thus a variety of parameterisations (refer to Sec. 2.2.1 for meaning) can be utilised to describe a single process (Tian 2006). Consequently, there are inter-model differences in the description of key processes related to iron (Tagliabue *et al.* 2016). Therefore, there is a need to understand the implications of different iron parameterisations in effecting biogeochemical model outputs.

1.2 Scope of project

The book work will focus on the inorganic iron parameterisations of two biogeochemical models. The idea is to test the various parameterisations and understand their impact on the model outputs. However, the work does not intend to validate the respective iron parameterisations, instead, it aims to showcase the different biogeochemical responses that occur due to the choice of iron parameterisations. In no way is this work an exhaustive assessment of all the inorganic iron parameterisations that are used in current biogeochemical models. However, the chosen formalisms are commonly employed in most biogeochemical models capable of representing iron (Tagliabue *et al.* 2016).

1.3 Aims and objectives

The aim of the book is to compare two sets of iron parameterisations within a background numerical model to understand the implications of the choice of iron parameterisations have on the functioning of a biogeochemical model. Focus will be given to the parameterisations of scavenging (refer to Sec. 2.1.1 and 2.2.3) as it is an important process in the iron cycle and is poorly constrained in current ocean biogeochemical models (Tagliabue *et al.* 2016, Yao *et al.* 2019) which reflects a lack of knowledge and scientific consensus on the process itself. To accomplish this, the second chapter presents a literature review which focusses on the description of the iron cycle as well as exploring the various iron parameterisations employed in current biogeochemical models. The third chapter is the methodology and it describes the background biogeochemical model, the various numerical experiments and the statistical test utilised on the model outputs. The results and discussion chapters will explain and interrogate the modelling repercussions for the choice of iron parameterisations while the conclusion will answer two important questions: 1) Is the choice of iron parameterisations significant when running a biogeochemical model? 2) Can 0D models be used as spaces to test and understand accurately the modelling repercussions for different parameterisations?

2. Literature review

Iron is one of the most commonly studied trace metals as it exerts a significant influence on ocean productivity, carbon sequestration as well as modulating atmospheric CO_2 concentrations (Boyd & Ellwood 2010). As iron is such a vital nutrient for biogeochemical processes it is often included as a variable in biogeochemical models (Tagliabue *et al.* 2017). The literature review is divided into two main sections: biogeochemistry of iron and biogeochemical modelling. The first section introduces the iron cycle (Gledhill & Buck 2012) and explores the major components of: sources of iron, biogeochemical processes and organic ligands. This is necessary in order to highlight the variety of biological, physical and chemical interactions that have to be modelled and parameterised in biogeochemical models. Consequently, the second section introduces the concept of a biogeochemical model before addressing the various parameterisations and mathematical formalisms utilised to describe key biogeochemical processes related to iron such as scavenging and complexation to organic ligand.

2.1 Biogeochemistry of iron

2.1.1 Iron Cycle

The earth system comprises of chemical, physical, biological and human influences that manifest themselves as multiple non-linear responses and linkages between the different components (Jickells *et al.* 2005). The iron cycle is one such process (Fig. 1) and it involves the complex interactions between lithogenic inputs, dissolution, precipitation, scavenging, biological uptake, remineralisation and sedimentation dynamics (Gledhill & Buck 2012). Martin & Fitzwater (1988) showcased iron to be an important limiting nutrient for phytoplankton growth in the High Nutrient Low Chlorophyll (HNLC) regions of the North-east Pacific (NEP) as well as the Southern Ocean (SO) (Martin, Gordon & Fitzwater 1991). It has been estimated that HNLC regions constitute 25% of the world's ocean (Boyd & Ellwood 2010) and represent places of potential CO_2 drawdown. Martin & Fitzwater (1988) only did bottle iron-enrichment studies to showcase the increased utilization of excess nitrate in HNLC regions and noted the link between increased iron supply and elevated CO_2 drawdown known as the *iron hypothesis*.

Consequently, the *iron hypothesis* spurred on the need for mesoscale iron fertilisation experiments such as the IronEx project in the east Equatorial Pacific (EP) (Martin *et al.* 1994; Coale *et al.* 1996; Landry *et al.* 2000) and the Southern Ocean Iron-Release Experiment (SOIREE) (Boyd *et al.* 2000) to assess the viability of HNLC regions as being places to sequester CO_2 from the atmosphere. Both IronEx and SOIREE corroborated the ideas of Martin *et al.* (1991) with elevated phytoplankton growth and increased chlorophyll concentrations occurring as a result of iron fertilisation (Martin *et al.* 1994; Coale *et al.* 1996; Boyd *et al.* 2000; Landry *et al.* 2000). However, Boyd *et al.* (2000) cautioned about the viability of iron-enrichment leading to elevated carbon sequestration. A follow up model study by Aumont & Bopp (2006) stressed that iron fertilisation was not the solution for stemming the rise in atmospheric CO_2 concentrations, citing the large uncertainties relating to the fate of sequestered carbon as a major barrier.

Sources of iron

As first noted by Martin & Fitzwater (1988), the supply of iron is a limiting factor to phytoplankton growth over most areas of the ocean. Exogenous iron reaches the ocean in three major ways (Fig. 1): river and fluvial deposits, hydrothermal vents and aeolian deposition (Jickells *et al.* 2005; Boyd & Ellwood 2010). River and fluvial deposits as well as sedimentary and glacial particulate iron only nourish the coastal and near coastal environments (Jickells *et al.* 2005, Rijkenberg *et al.* 2014). Consequently, iron is found at 100 to 1000 times greater concentrations within coastal environments compared to the open ocean (Sunda & Huntsman 1995). This strong horizontal gradient in concentration has a profound effect on the respective biological communities. Sunda & Huntsman (1995) showed that oceanic phytoplankton species tended to be smaller in cell size, had decreased demand for iron containing enzymes and had lower growth requirements for cellular iron when compared to coastal phytoplankton. However, the high iron availability in coastal waters could permit luxury uptake; which is the ability to take up and store iron in excess levels needed for immediate metabolic requirements. Therefore, it is hypothesised that luxury uptake would be advantageous to coastal species where iron concentrations are high but temporally variable (Sunda & Huntsman 1995).

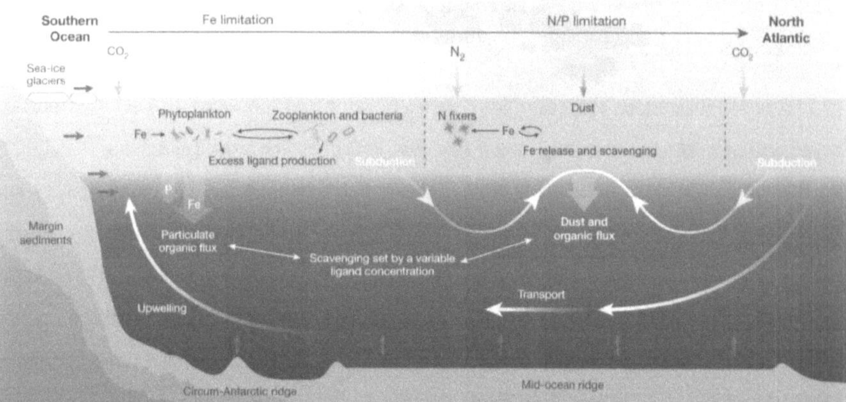

Figure 1: The iron-cycle involves the interplay of multiple biotic and abiotic components that manifest themselves within the land, atmosphere and ocean. A transect along the Southern Ocean to the North Atlantic reveals the spatial disparities within the iron cycle between these two regions. Most pertinent being the high aerosol flux received in the North Atlantic in contrast to the Southern Ocean; resulting in the North Atlantic being nitrogen/phosphorous limited. In addition to aerosols, hydrothermal vents and fluvial deposits act as sources of iron to the ocean. Excess ligand production occurs in the Southern Ocean owing to the high biological productivity. Dust particles play a part in the scavenging of bioavailable iron while bacteria remineralise the particulate organic iron. (Source: Tagliabue *et al.* 2017)

Hydrothermal vents have their iron inputs rapidly dissipated at depth but can act as an important source for the deep ocean (Tagliabue *et al.* 2010), especially along the Mid-Atlantic Ridge (Lough *et al.* 2019). However, the most important exogenous source of iron to the open ocean is aeolian dust (Jickells *et al.* 2005). Hyper-arid regions comprise nearly a third of global land area and are the major sources for dust particles. Desert dust comprises of μm sized particles that can have lifetimes of hours to weeks in the atmosphere, allowing them to be transported great distances (Duce & Tindale 1991). However, dust particle production, transportation and deposition are a function of multiple non-linear factors such as climate, topography and vegetation cover (Duce & Tindale 1991; Jickells *et al.* 2005). There is a large spatial disparity between ocean basins in terms of aerosol fluxes with regions such as the North Atlantic receiving large quantities of dust as a result of the Sahara (Jickells *et al.* 2005;

Anderson *et al.* 2016) while the SO is characterised by a low aerosol flux (Martin & Fitzwater 1988; Boyd & Ellwood 2010).

In addition to exogenous sources of iron, vertical mixing and upwelling also act as important sources of new iron to the photic zone (Falkowski, Barber & Smetacek 1998; Fung et al. 2000). Furthermore, iron can also be made biologically available to phytoplankton through recycling (regenerated production) in the upper layers of the ocean. Fung *et al.* (2000) studied the cycling dynamics of iron in the upper-ocean and explored the recycling efficiency of iron in the major HNLC regions of the SO, EP and NEP as well as the high atmospheric deposition region of the North-west Pacific (NWP). Using numerical simulations, Fung *et al.* (2000) calculated the recycled iron flux as being the difference between fixed iron and supplied iron, noting uncertainties relating to preferential consumption of recycled or new iron for phytoplankton. Consequently, for regions such as the SO, which have a small aeolian input, >95% (Fung *et al.* 2000) of production had to be supported by regenerated iron. Whereas for the NWP, the high atmospheric deposition precluded the need to regenerate large quantities of iron to sustain production. The EP differed to the SO and NEP as the EP is a prominent upwelling region which results in an additional source of iron for the upper-ocean, ensuring lower levels (<30%) of regenerated production (Fung *et al.* 2000).

Therefore, exogenous sources of iron as well as upwelled and regenerated production of iron act as important sources for phytoplankton. Once iron is made available, it is transformed by multiple biogeochemical processes.

Biogeochemical processes

Iron is the fourth most abundant element in the Earth's crust (Falkowski, *et al.* 1998) and is an essential micronutrient for phytoplankton as it has been linked to key processes such as photosynthesis, respiration and nitrogen fixation (Morel & Price 2003). However, iron occurs at concentrations less than 1 nM in most surface oceanic waters (Martin *et al.* 1991) due to iron's low solubility in seawater (Falkowski *et al.* 1998) and rapid scavenging, utilisation and complexation (Rue & Bruland 1995) by biotic and abiotic mechanisms (Boyd & Ellwood 2010). Therefore, from a biogeochemical context, the key flux to the ocean is not particulate iron but rather soluble or dissolved iron.

Dissolved iron is unique among nutrients as it has a short residence time, 100-200 years, (Johnson, Gordan & Coale 1997) when compared to the time scale of the thermohaline circulation (1000 years). As exhibited by Martin *et al.* (1991), dissolved iron has a nutrient-like profile (Fig. 2) indicative of a biological influence. Subsequently, low dissolved iron concentrations occur in the surface waters as a result of uptake by phytoplankton and increasing concentrations are observed with depth as a consequence of remineralisation, mainly by heterotrophic bacteria (Morel & Price 2003). However, Johnson *et al.* (1997) noted that this simple uptake and remineralisation scheme cannot account for the rapid formation of a nutrient-like profile (Fig. 2) and suggested that iron concentrations were maintained by organic ligands that complexed iron which limited scavenging. Consequently, the complexation to organic iron-binding ligands plays a significant role in controlling the concentration of dissolved iron in the ocean (Gledhill & Buck 2012).

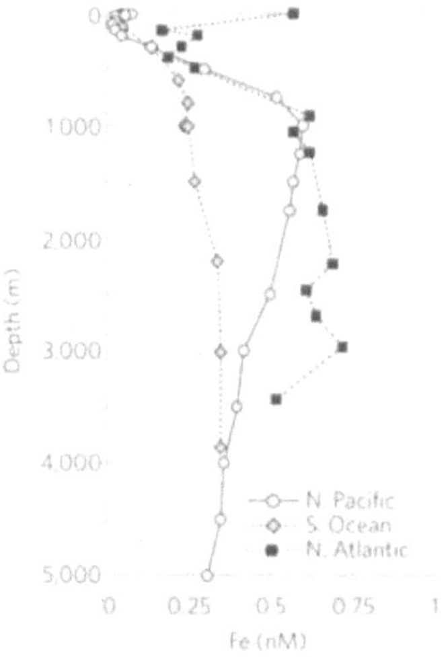

Figure 2: The dissolved iron profiles for the North Atlantic, North East Pacific and Southern Ocean observe low concentrations of dissolved iron in the surface waters due to uptake by phytoplankton while increasing concentrations with depth are associated with remineralisation. (Source: Boyd & Ellwood 2010)

Organic ligands

Dissolved iron (< 0.4 μm) can take on multiple physio-chemical forms which include: Fe(III), Fe(II), colloidal (0.02-0.4 μm), truly soluble (< 0.02 μm), organically complexed iron and inorganic iron (Boyd & Ellwood 2010). However, almost 99% of dissolved iron is complexed to organic iron-binding ligands (Fig. 3b) (Rue & Bruland 1995) which act to buffer dissolved iron concentrations in seawater, limiting hydrolysis, precipitation and particle scavenging (Gledhill & Buck 2012). Organic ligands are molecules that can bind to, and form stable complexes with trace metals in the aquatic dissolved phase (Völker & Tagliabue 2015) and they are an important component in the Dissolved Organic Carbon (DOC) pool as they act to increase the solubility of iron and hence the availability to phytoplankton (Hassler, van den Berg & Boyd 2017). Traditionally, electrochemical detection methods distinguish between two types of organic ligands, a 'strong' binding ligand (L1) and a 'weaker' ligand (L2) which have different affinities for iron (Hunter & Boyd 2007). However, the production of organics ligands by various biological processes lends itself to the existence of multiple species of ligands.

Over most regions of the ocean, organic ligand concentrations exceed that of dissolved iron on average by 1 nM (Gledhill & Buck 2012). In the upper ocean, bacteria can produce L1 binding siderophores, which is an iron-chelating compound, to acquire iron (Tortell *et al.* 1999; Barbeau *et al.* 2001) while phytoplankton can release several L2 ligands such as domoic acid, saccharides and exopolymetric substances (EPS) (Hassler *et al.* 2017). In addition, L2 ligands are produced by passive processes linked to exudate or remineralisation of cellular debris (Gledhill & Buck 2012) which leaves behind less labile dissolved organic matter as a humic like material. (Hassler *et al.* 2017). Furthermore, Fig. (3b) shows that iron-binding ligands are found in most iron sources such as: glacial-ice, dust and hydrothermal vents, highlighting the tight coupling between iron and its complexing pair.

Iron-binding ligands are not likely to be long-lived on the scale of the thermohaline circulation as they are affected by bacterial as well as photochemical degradation in the surface layers (Barbeau *et al.* 2001; Hunter & Boyd 2007) and aggregation onto sinking particles (Völker & Tagliabue 2015) (Fig. 3b). Consequently, the interplay of sources and sinks for ligands result in the surface waters (upper 100 m) having the greatest and most variable concentration of ligands (Fig. 3A), with a peak occurring around the subsurface chlorophyll maximum, corresponding to high biomass accumulation (Gledhill & Buck 2012). However, processes such as remineralisation and photochemistry may be dualistic as being sources and or loss terms for organic ligand, showcasing the complexity of ligand cycling (Hassler *et al.* 2017).

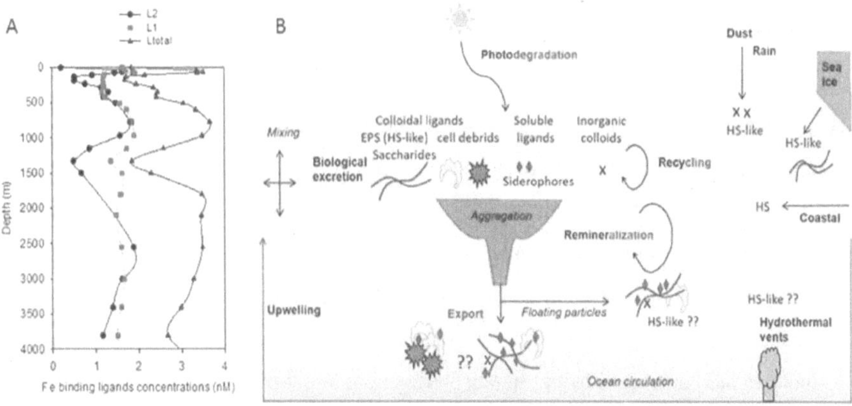

Figure 3: ***A****, vertical profile of organic ligands taken in the North Atlantic showing the distribution of strong (L1), weak (L2) and total (Ltotal) concentrations.* ***B****, illustration of the cycling of ligands in the ocean highlighting the various sources (black bold) and loss terms (red bold) as well as the flow of matter indicated by the arrows. Humics are subdivided into two categories: terrestrial based (HS) and marine based (HS-like). While siderophores and various biological excretions (EPS) are shown in association with the accumulation of organic matter.* (Source: Hassler *et al.* 2017)

2.1.2 Summary

A dynamic relationship exists between ocean biota and iron as phytoplankton and bacteria control the chemistry and cycling of iron while iron controls the growth of the organisms, and in turn, the cycling of other major nutrients such as carbon and nitrogen. An appreciation of the iron cycle in the world's oceans requires the integration of diverse fields of knowledge, ranging from global geochemistry to cellular-scale genetics. Consequently, the iron cycle is incredibly complex as it involves the interplay of numerous physical and biological components resulting in multiple uncertainties. Although decades of research have improved our understanding of the iron cycle, quantification of the fluxes of iron to the ocean (Jickells *et al.* 2005) as well as understanding the spectrum of iron-binding ligands and their interactions with different iron species (Gledhill & Buck 2012) are required. Therefore, an integrated view of the iron cycle places additional challenges on biogeochemical models that are utilised for hypothesis testing and climate change projections.

2.2 Biogeochemical modelling of the iron cycle

2.2.1 What is a biogeochemical model?

At present, a quantitative appreciation of marine ecosystems requires two major research components: marine biology and physical oceanography. In contrast to physical oceanography, where the basic hydrodynamic equations have their origins in mathematical formulas of fundamental laws, biological models cannot be derived from first principles because ecological dynamics play a significant role (Fennel & Neumann 2001; Fulton, Smith & Johnson 2003; Tian 2006). Therefore, a biological model consists of a number of PDEs that are formulated mathematically by translating observations into formulas that are consistent with ecological principles. In doing so, a spectrum of biogeochemical models that describe a marine ecosystem exist and they encompass a variety of parameterisations, spatial as well as process resolutions.

Parameterisations

A parameterisation refers to a formula or set of formulas that are used to describe and quantify a controlling process (Fennel & Neumann 2001); however, some authors use the word 'model formalism' synonymously. With relevance to marine ecosystems, this basically manifests itself as the description of gains and losses in a state variable. Therefore, let X_i be a state variable that can be described in terms of concentration per unit volume of water and thus the change in X_i by a marine ecosystem can be expressed in Eq. (1.1):

$$\frac{d}{dt}X_i = (gain(X_j) - loss(X_j))X_i + Q^{external} \qquad (1.1)$$

where the processes that control the change in X_i may depend on another state variable X_j as well as external variables which can be both physical and/or biological. Consequently, processes such as the uptake of nutrients or ingestion are representative of gain terms while cell lysis, respiration and egestion are loss terms. However, there is no generally accepted set of equations to model a marine ecosystem and thus a variety of parametrisations can be utilised to describe a single controlling process (Pereira, Duarte & Norro 2006; Tian 2006). In addition, different biogeochemical models may model the same processes, but with different degrees of detail. Inherently, the degree of detail for a process is predicated on its importance as well as available knowledge (Pereira *et al.* 2006).

A common example is phytoplankton growth which can be parameterised with three different models: Monod, quota and mechanistic (Flynn 2003). The Monod model is the simplest as it relates the growth of phytoplankton as a function of the external dissolved concentration of the limiting nutrient (Sommer 1991). Therefore, if X is the limiting nutrient and K_X is the half-saturation constant for growth, then the Monod model can be written in Eq. (1.2) as:

$$\mu = \mu_{max} \frac{X}{X+K_X} \tag{1.2}$$

Where μ_{max} is the maximum specific growth rate for a set of temperature and light conditions while μ is the actual growth rate. Though simple, the Monod model is only suited to steady-state simulations and struggles to represent systems with multi-nutrient interactions. In addition, the Monod model does not permit phytoplankton to utilise their internal quotas of nutrients in the absence of external concentrations (Flynn 2003).

Consequently, the quota model improves upon the Monod by relating phytoplankton growth to the internal availability of nutrients (Flynn 2003). Thus the quota model is an intrinsic function, where phytoplankton growth is a function of internal nutrient content and this in turn is a function of the external nutrient concentration in the ocean. Typically, the quota model uses nutrient ratios in terms of carbon (C) and two of the most commonly used quotas models are the Droop (Eq. 1.3) and Caperon & Meyer (Eq. 1.4) (Sommer 1991).

$$C_\mu = \mu_{mX} \frac{X_C - X_{C_0}}{X_C} \tag{1.3}$$

$$C_\mu = \mu_{mX} \frac{X_C - X_{C_0}}{X_C - X_{C_0} + K_X} \tag{1.4}$$

Therefore, let C_μ be the carbon related growth rate while X_C is the nutrient:C quota and X_{C_0} is the minimum quota at which phytoplankton can survive. In addition, K_X is a curve fitting constant and μ_{mX} is the maximum specific growth rate when using nutrient X as substrate.

The trend in biogeochemical modelling is to develop more mechanistic descriptions of biogeochemical processes based on physiological and biological dynamics rather than relying on empirically derived functions as seen in Eq. (1.2-4) (McDonald & Urban 2010). However,

Crout, Taritano & Wood (2009) state that mechanistic models are likely to be less detailed than the system they seek to describe which inevitably results in them becoming over-parameterised. In addition, Flynn (2003) notes that though mechanistic models attempt to include more biologically meaningful interactions, we lack the necessary knowledge to construct them.

Consequently, as biogeochemical parameterisations become more complex, greater uncertainty is incurred in model formalisms due to the increase in the number of parameter values. Though some parameter values can be constrained, insufficient data exists to deal with all parameters, even in simple models (Ward *et al.* 2010). Incidentally, efforts can be made to reduce model uncertainty by reducing the number of model variables (Crout *et al.* 2009) or adopt parameter optimization techniques (Annan *et al.* 2005; Ward *et al.* 2010) that assign optimal values. Parameter optimisation is advantageous as it can reduce model error, relative to observational data, compared to hand-tuned models (Yao *et al.* 2019) and from a methodological perspective, assigning optimal parameter values makes inter-model comparisons more fair as the true difference in model behaviour can be attributed to model structure rather than the relevant parameter values (Ward *et al.* 2010).

Iron parameterisations

In current global biogeochemical models, the iron cycle is usually resolved into a phytoplankton, dissolved and particulate component (Fig. 4) (Moore *et al.* 2002; Vichi, Pinardi & Masina 2007b; Aumont *et al.* 2015). Typically, the dissolved component is seen as completely bioavailable to phytoplankton (Vichi *et al.* 2007b) while the particulate component can be divided into separate species along particle size (Moore *et al.* 2002; Aumont *et al.* 2015) or dissolution state (Vichi *et al.* 2007b). Most biogeochemical models consider iron as an essential nutrient for phytoplankton growth (Tagliabue *et al.* 2016) but unlike carbon, nitrogen and phosphorous; iron is typically added as a separate multi-nutrient limitation term within phytoplankton (Vichi *et al.* 2007b).

In representing the iron cycle, biogeochemical models must parameterise the major processes of: uptake by phytoplankton, remineralisation and scavenging (Fig. 4). Dissolved iron is typically modelled using the quota model for phytoplankton growth (Vichi *et al.* 2007b; Aumont *et al.* 2015) while the process of remineralisation is generally coupled to the dynamics of organic matter by multiplying the concentration of particulate carbon by a fixed iron:carbon ratio and this early approach to iron remineralisation has been employed by several other authors (Archer & Johnson 2000, Parekh, Follows & Boyle 2004). Even in more advanced biogeochemical models, remineralisation is generalised as a linear process (Moore *et al.* 2002; Vichi *et al.* 2007b; Aumont *et al.* 2015). Multiple parameterisations can be utilised to model the scavenging of iron (Archer & Johnson 2000; Parekh *et al.* 2004) and the process is one of the least constrained in the biogeochemical modelling of the iron cycle (Tagliabue *et al.* 2016, Yao *et al.* 2019). Consequently, Sec. (2.2.2) explores the current generation of global biogeochemical models, with Tagliabue *et al.* (2016) noting that the inter-model differences in the scavenging parameterisations as well as the dynamics of organic ligands play a significant role in dictating the concentration of dissolved iron in the global ocean. As the scavenging regime exerts a significant influence on the concentration of dissolved iron, the various scavenging parameterisations employed in current biogeochemical models are explored in detail in Sec. (2.2.3).

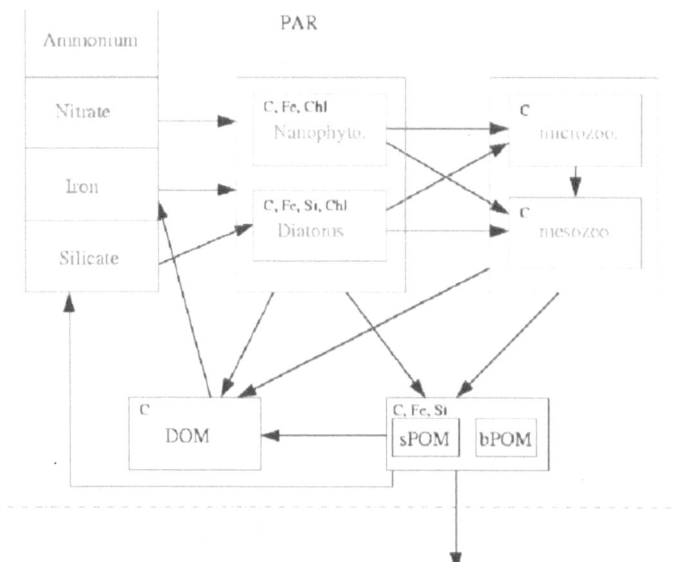

Figure 4: Flow diagram of the PISCES biogeochemical model. Arrows represent the flow of matter and energy between the state variables which is governed by various processes such as uptake, lysis, remineralisation and scavenging. (Source: Aumont *et al.* 2015)

Spatial resolution

Biogeochemical models can be run as simple box models or be coupled to 3D global-circulation models. Box models assume instantaneous homogeneity of all state variables in the given spatial enclosure. According to Fennel & Neumann (2001), they are useful for theoretical studies with multiple state variables and to explore as well as identify key processes that drive a biogeochemical system. Whereas coupled 3D ocean models are used extensively in modelling large scale dynamics such as the large scale distribution of plankton (McKiver *et al.* 2015). In addition, coupled models are used as forecasting tools, most notably being projections of change relating to rising atmospheric CO_2 concentrations (Bonan & Doney 2018).

A general trend in ecosystem modelling is to improve spatial resolution by shifting from box and low resolution models (Ménesguen *et al.* 2007) to fine grid, high resolution 3D models (McKiver *et al.* 2015). However, conflicting arguments exist regarding the use of finer resolution models as being solutions to resolve discrepancies between coupled biogeochemical model outputs and observational data. Though increasing the complexity of a biogeochemical model may result in improved representations of the ecosystem dynamics, under-resolving the physical processes that have a direct influence on the biogeochemical system will also significantly affect model outputs. McKiver *et al.* (2015) used a coupled biogeochemical model to investigate the impact of mesoscale processes on the global marine biogeochemical system by utilising a low (2°) and high (¼°) resolution model. By improving the vertical and spatial resolution, McKiver *et al.* (2015) saw improvements in representing the chlorophyll dynamics of: coastal regions, subtropical gyres and the SO due to the high-resolution model better resolving the vertical physics.

Ménesguen *et al.* (2007) had a similar methodology to McKiver *et al.* (2015) where a biogeochemical model of pelagic primary production was coupled to a physical ocean model of differing spatial resolutions. Using a two and three-layered box model as well as a fine resolution 3D model, Ménesguen *et al.* (2007) assessed their differing capabilities in representing the biogeochemistry of the English Channel. With relevance to annual primary productivity, both the box and 3D models showed similar results; with all the models having an inadequate phasing of the chlorophyll maximum. Therefore, Ménesguen *et al.* (2007) suggested that improved representations of the biogeochemical system would only be accomplished with better biogeochemical parameterisations rather than more refined spatial resolutions.

Consequently, the principle of parsimony would require a balance between the complexity of the biogeochemical parameterisations and the spatial resolution. Therefore, for biogeochemical tracers that have a global valence, such as iron, their ultimate implementation should be within coupled 3D models. However, owing to large uncertainties relating to the parameterisations of the iron cycle in biogeochemical models (Yao *et al.* 2019), box or low resolution models can provide an ideal environment for testing and refining biogeochemical processes before they are embedded in fine resolution 3D simulations.

Process resolution

A marine ecosystem can be modelled with varying degrees of complexity by altering the number of state variables as well as the detail of the biogeochemical parameterisations used to model individual processes. Consequently, the structural complexity of a model will influence the resolution of the processes that wish to be studied and this is dependent on the scientific question that needs to be addressed.

The state variables of nutrients, phytoplankton, zooplankton, dissolved and particulate organic matter in themselves can have varying degrees of process resolutions in relation to the required modelling scenario. Nutrients can be envisaged as a single limiting nutrient or consist of a host of macronutrients (nitrate, phosphate and silicate) as well as micronutrients such as iron (Flynn 2003). In addition, phytoplankton can be represented as a single bulk biomass for showing general patterns of biological activity. In more advanced models, phytoplankton can be resolved into separate functional groups corresponding to: diatoms, flagellates and cyanobacteria to account for their differing uptake rates, sinking speeds and nutrient preferences (Vichi *et al.* 2007b). Similarly, zooplankton can simply be parameterised as a grazing pressure term on phytoplankton or be modelled as a separate functional groups with different levels of predation rates. Pegged to phytoplankton-nutrient dynamics, detritus can be partitioned in correspondence with the phytoplankton variables or be considered as a single bulk variable (Fennel & Neumann 2001).

Intuitively, the increase in the structural complexity of a model should lead to a reduction in model error with observational data. Kriest *et al.* (2010) used a biogeochemical model based on phosphorous and set-up a hierarchy of models with increasing structural complexity. The simplest only considered the nutrient phosphate while the most complex included the interactions of particulate and dissolved organic phosphate, phytoplankton and zooplankton. However, Kriest *et al.* (2010) found that merely increasing the number of model components does not necessarily lead to improved correspondence with observational data. Following a similar methodology to Kriest *et al.* (2010), Yao *et al.* (2019) used a calibrated coupled biogeochemical model and adjusted the structural complexity of the iron module to investigate whether explicitly representing the processes of iron would lead to a reduction in model-data misfits. Three variants of an iron module were used, with the first explicitly resolving the iron cycle while the second considered iron limitation in primary productivity by utilising an iron mask of prescribed monthly concentrations of dissolved iron and the third variant did not represent the iron cycle. Yao *et al.* (2019) found that using an explicit module for the iron cycle lead to improvements in representing the distribution of macronutrients (phosphate, nitrate and silicate) as well as oxygen in the global ocean.

Therefore, Yao *et al.* (2019) concluded that increasing the process resolution of iron in a biogeochemical model was important as it lead to improved representations of global biogeochemical nutrient cycles. Consequently, process resolution encompasses multiple facets of biogeochemical modelling and thus the choice of complexity regarding state variables and parameterisations must compliment the system being modelled.

2.2.2 Current state of biogeochemical models in representing iron dynamics

The iron cycle plays an important role in ocean biogeochemistry and received extensive academic attention in the 1980's; however, it was not until the early 2000's that iron was included as a variable in major biogeochemical models (Tagliabue *et al.* 2017). At present, a host of biogeochemical models exist such as: Biogeochemical Flux Model (BFM) (Vichi *et al.* 2007b), Biogeochemical Elemental Cycling (BEC) (Moore *et al.* 2002; Moore, Doney, & Lindsay 2004) and the Pelagic Interactions Scheme for Carbon and Ecosystem Studies (PISCES) (Aumont *et al.* 2015) (Fig. 4) to list a few. All of these models simulate marine biological productivity and describe the cycling of major nutrients such as: C, P, N, Si and Fe.

In comparison to the early iron models of Archer & Johnson (2000) and Parekh *et al.* (2004), the current suite of biogeochemical models has integrated iron into the living functional groups. They have also improved the process resolution of the iron cycle by incorporating multiple iron sources, including: riverine, dust and hydrothermal (Tagliabue *et al.* 2016) as well as refining the parameterisations of key processes such as scavenging (Moore & Braucher 2008) and ligand dynamics (Tagliabue & Völker 2011, Völker & Tagliabue 2015) (refer to Sec. 2.2.3).

<u>Inter-comparison of biogeochemical models that represent dissolved iron</u>

The increased observational data provided through the GEOTRACES programme (Mawji *et al.* 2015; Schlitzer *et al.* 2018) has allowed a more rigorous assessment of the current suite of biogeochemical models that are capable of representing the cycling of iron. Subsequently, Tagliabue *et al.* (2016) conducted a comparison of 13 major biogeochemical models (Fig. 5) that represent iron dynamics with the 2015 GEOTRACES data (Mawji *et al.* 2015) known as the Iron Model Inter-comparison Project (FeMIP). Fig. (5) highlights the inter-model differences in representing the distribution of surface dissolved iron with models such as the BFM and BLING having significantly lower concentrations of dissolved iron in the polar regions while TOPAZ, MEDUSA1 and MEDUSA2 have higher dissolved iron concentrations in the ocean gyres.

The study revealed that contemporary models contain a greater array of iron sources, with most including a dust and sediment source; but fewer models having hydrothermal and river inputs. Even for a given source, there was still significant inter-model differences in the strength of the iron flux, the most prevalent being the dust source (Tagliabue *et al.* 2016). However, though the range of total iron inputs between the various FeMIP models varied substantially (66.9 ± 67.1 Gmol Fe yr^{-1}), the mean dissolved iron concentration was similar (0.58 ± 0.14 nM) (Tagliabue *et al.* 2016). From Fig. (5) it is clear that the current suite of FeMIP models struggle to replicate the observational patterns of dissolved iron in the surface, owing to the intricacies of the iron cycle and the inherit knowledge gaps therein. Therefore, Tagliabue *et al.* (2016) attributed the similar mean dissolved iron content to the various scavenging regimes employed in the respective models with most using the formalism of Parekh *et al.* (2004) except for BFM, COBALT, BEC, MEDUSA1 and MEDUSA2. However, Tagliabue *et al.* (2016) acknowledged that the various iron parameterisations for each respective model were not evaluated and instead their coupled physical-biogeochemical framework was compared. Consequently, there is a necessity to evaluate and constrain the various scavenging parameterisations and rates to improve the comparability as well as the functionality of biogeochemical models in the representation of ocean iron dynamics.

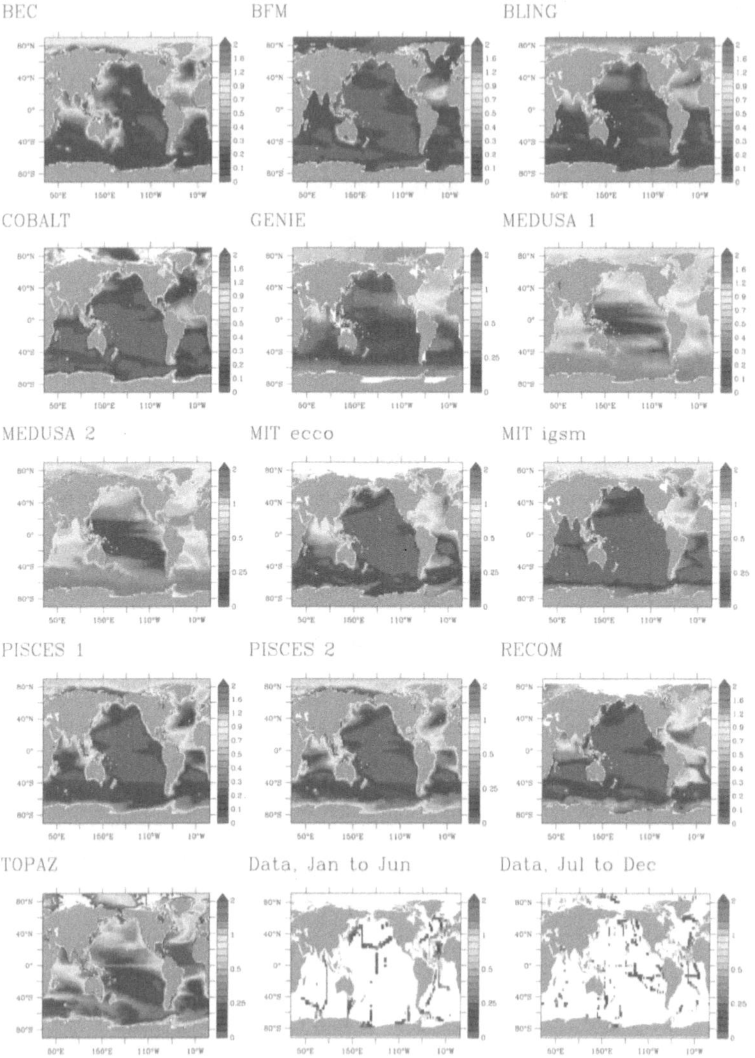

Figure 5: Comparison of annual mean surface dissolved iron concentration (nM) across 13 biogeochemical models with cruise data from the GEOTRACES programme. (Source Tagliabue *et al.* 2016)

Furthermore, there is a limit to which more advanced parametrisations can reduce model error without any fundamental changes in the understanding of the biogeochemistry of iron. Subsequently, more complexity does not necessarily lead to better results as increasing the complexity of a system increases the uncertainty by adding more degrees of freedom. Although multiple uncertainties exist in the cycling of iron, biogeochemical models that include the iron cycle will produce more realistic outputs than models without.

2.2.3 Different iron parameterisations for scavenging

Iron bioavailability is altered by several key processes: scavenging by inorganic and organic particles, remineralisation and biological uptake. Consequently, the parameterisations of these various processes have evolved in tandem with increased observational measurements. Tagliabue *et al.* (2016) as well as Yao *et al.* (2019) have showed that the iron scavenging parameters are not well constrained and significant inter-model differences exist in the parametrisation of the scavenging regime. Consequently, this section will focus on the process of scavenging and explore two different complexation formulations: a constant ligand model (Johnson *et al.* 1997, Archer & Johnson 2000) and a free scavenging model (Rue & Bruland 1995, Parekh *et al.* 2004). In addition, focus will be given to how current biogeochemical models have made improvements regarding scavenging rates and ligand dynamics.

Constant ligand model

Iron differs from other scavenged metals such as lead, aluminium and thorium because iron is utilised by biota for growth (Gledhill & Buck 2012). Johnson *et al.* (1997) utilised sample data from the North and South Pacific, SO as well as the North Atlantic to quantify the processes that controlled the distribution of dissolved iron by means of a numerical model. Based on the consistency of deep dissolved iron concentrations, Johnson *et al.* (1997) suggested that the removal rates of dissolved iron must decrease below concentrations of 0.6 nM and this was maintained by a mechanism of strong iron-binding ligands (Rue & Bruland 1995). Therefore, Johnson *et al.* (1997) parameterised scavenging J_{Fe} as a first-order process (Eq. 1.5):

$$J_{Fe} = k_{Fe}[Fe_T - L_T] \tag{1.5}$$

Where Fe_T is the total dissolved iron concentration and L_T is the total ligand concentration (Archer & Johnson 2000), set to 0.6 nM. A switch function was employed whereby the scavenging rate constant $k_{Fe} = 0$ when $Fe \leq L_T$ and the value of k_{Fe} was treated as an adjustable parameter for dissolved iron concentrations $> L_T$. After calibrating the scavenging rate, Johnson *et al.* (1997) used their model to reproduce vertical profiles of dissolved iron from stations in the North Atlantic, SO as well as the Pacific. To accomplish this, Johnson *et al.* (1997) prescribed the export flux of carbon in the various regions to account for the differing levels of biological productivity. Consequently, the model was able to resolve many of the open ocean stations with careful calibration of the carbon export rate but struggled to represent coastal systems which are influenced by additional iron sources such as riverine deposits.

A follow up model study by Archer & Johnson (2000) sought to contextualize the iron parameterisations of Johnson *et al.* (1997) in a global circulation model where primary production was limited by the availability of phosphate and iron using a Monod approach. In addition, the formalism of Johnson *et al.* (1997) was applied with two ligands, a weak and strong, using the depth-dependent concentration of the respective iron-binding ligands measured by Rue & Bruland (1995). Unlike the single ligand model of Johnson *et al.* (1997), scavenging was permissible with the two-ligand model at concentrations below 0.6 nM. The study did not seek to validate the various scavenging parameterisations and instead showcased the global distribution of dissolved iron using a coupled ocean model; noting excess dissolved iron concentrations near coastal and upwelling regions and the deficiency of dissolved iron in the SO.

Free scavenging model

The bulk concentration of dissolved iron exists in a non-reactive colloidal state due to the binding to organic ligands (Morel & Price 2003). To understand how organic complexation by ligands influenced the speciation of dissolved iron, Rue & Bruland (1995) constructed a theoretical mass balance equation (Eq. 1.6):

$$[Fe_T] = [Fe'] + [Fe_L] \tag{1.6}$$

In Eq. (1.6), Fe_T is the total concentration of dissolved iron consisting of: Fe' which represents the sum of all inorganic species while Fe_L represents the organically complexed fraction. Using this model, Rue & Bruland (1995) showed that 99.9% of dissolved iron exists in a complexed form with a marginal fraction existing as uncomplexed or 'free iron'. Embedding the study in

a six-box ocean model with the tracer elements of dissolved iron, particulate iron and phosphate, Parekh *et al.* (2004) sought to improve upon the work of Archer & Johnson (2000) by including a more mechanistic description of iron complexation to organic ligands by incorporating the idea of free iron. Where Archer & Johnson (2000) used L_T equivalent to 0.6 nM, Parekh *et al.* (2004) assigned L_T being equal to the sum of complexed iron and uncomplexed ligands L' (Eq. 1.7):

$$L_T = [Fe_L] + [L'] \qquad (1.7)$$

Using Eq. (1.6) and (1.7), Parekh *et al.* (2004) assumed that only the free form of iron was susceptible to scavenging:

$$J_{Fe} = k_{Fe}Fe' \qquad (1.8)$$

The parameterisations of Parekh *et al.* (2004) allow for the representation of free ligands as well as highlighting the inverse relationship between Fe_T and L_T. Therefore, the presence of a strong binding ligand ensures $Fe_T \approx FeL$ which is the limit modelled in Archer & Johnson (2000). However, Fe_T is highly sensitive to the choice of the L_T which means the scavenging constant k_{Fe} must be increased significantly to prevent elevated concentrations of Fe_T. In applying the scavenging parameterisation, Parekh *et al.* (2004) noted improved reproduction of the deep dissolved concentration of iron as well as the observed presence of uncomplexed organic ligands. Consequently, Parekh *et al.* (2004) suggested that the use of a weaker ligand and a greater total ligand concentration was more appropriate in reproducing the broad patterns of dissolved iron.

Therefore, the two models of Johnson *et al.* (1997) and Parekh *et al.* (2004) both acknowledge the importance of complexation by ligands in controlling the scavenging rate of dissolved iron. However, both parameterisations are still employed in sophisticated biogeochemical models; highlighting the variability present in the iron modelling community.

Improved process resolution of iron dynamics

Numerical models typically have parameters and constants that need to be assigned values in order to produce results (Rykiel 1996). The early models of Archer & Johnson (2000) and

Parekh *et al.* (2004) treated organic ligand concentrations (L_T) and the scavenging rate constant (k_{Fe}) as adjustable parameters that were spatially calibrated to agree with observational measurements. However, increased observational data has shown that ligands are spatially variable (Tagliabue & Völker 2011) and that the process of scavenging can include the interaction of lithogenic and biogenic particles (Moore & Braucher 2008). Consequently, a more prognostic parameterisation of scavenging rates and ligand dynamics was required to include more biologically meaningful processes.

Scavenging rate constant

Typically, the scavenging rate of dissolved iron (k_{Fe} in Eq. 1.5 and 1.8) is parameterised as a constant value around 0.005 year^{-1} to represent an estimated residence time of 200 years in the ocean (Johnson *et al.* 1997). In attempting to include a prognostic appreciation of scavenging, Moore & Braucher (2008) altered the iron scavenging rate (Moore *et al.* 2004) in BEC (Eq. 1.9), which was determined by a base scavenging coefficient (k_b) and scaled by the sinking particle flux of Particulate Organic Carbon (POC) and mineral dust ($Dust$). Furthermore, k_{Fe} could be adjusted by multiplying it by a coefficient to account for different concentrations of dissolved iron.

$$k_{Fe} = k_b(POC + Dust) \quad (1.9)$$

Moore & Braucher (2008) changed the definition of the scavenging rate constant (Eq. 1.10) to include biogenic silica (bSi), calcium carbonate ($CaCO_3$) as well as arbitrarily scaling POC to represent a larger weight of the particulate organic carbon fraction.

$$k_{Fe} = k_b[(POC \times 6) + Dust + bSi + CaCO_3] \quad (1.10)$$

The parameterisations of Moore *et al.* (2004) and Moore & Braucher (2008) allow for variability in the scavenging rate, which is dictated by aeolian deposition of dust particles as well as biological activity. Consequently, this prognostic approach permits variable scavenging rates, acknowledging the spatial variance in biological activity and dust flux between regions such as the SO and North Atlantic. By making the scavenging rate constant a variable, Moore & Braucher (2008) improved the representation of dissolved iron, especially in low iron regions. In addition, improved correlations with observational data were present for the surface and deep ocean.

Ligand dynamics

Most biogeochemical models assume a constant iron-binding ligand concentration, fixed between 0.6-1 nM (Tagliabue & Völker 2011). However, ligands observe spatial-temporal variations in their concentrations, prompting a prognostic approach to ligands in numerical models. As ligands have a biological origin, Tagliabue & Völker (2011) related the total ligand concentration to vary as a function of total DOC (Eq. 1.11) based on the observational work of Wagener, Pulido-Villena & Guieu (2008).

$$L_T = (DOC \times 0.09) - 3.2 \tag{1.11}$$

In a follow up model study, Völker & Tagliabue (2015) included the prognostic ligand parameterisation in two biogeochemical models, PISCES and REcoM. Comparisons with observational data showed that a prognostic ligand parameterisation yielded more nutrient-like profiles for dissolved iron than the explicit ligand formalisms. However, the elevated ligand concentrations resulted in increased dissolved iron concentrations in non-iron limited regions such as the Atlantic and Indian Oceans (Völker & Tagliabue 2015). This was due in part to the low scavenging rates of uncomplexed iron; prompting the need to re-evaluate the scavenging rates in the respective models.

2.2.4 Final remarks

Biogeochemical models are an abstraction of the complex ecosystem processes and they have grown in sophistication as well as complexity in tandem with our knowledge of the biosphere. They can be run with varying temporal, spatial and process resolutions in accordance with the scientific question that needs to be addressed. Within biogeochemical models, the iron cycle can be described with an array of parameterisations. Consequently, the early iron models of Archer & Johnson (2000) and Parekh *et al.* (2004) showcased different conceptualisations of scavenging and ligand dynamics. Further study has been dedicated to improving the description of these processes by adopting a prognostic approach which has seen positive results in subsequent model studies. However, the current suite of FeMIP models still struggle to model the iron system.

Consequently, there is a limit to which more advanced parametrisations can reduce model error and uncertainty without any fundamental changes in the understanding of the biogeochemistry of iron. Increasing model complexity through the addition of more advanced parameterisations reduces model error relative to observational data by improving the process realism of the model. However, the addition of more complex parameterisations inherently increases the overall model uncertainty by increasing the degrees of freedom. Although multiple uncertainties exist in the cycling of iron, biogeochemical models that include the iron cycle will produce more realistic outputs than models without (Yao *et al.* 2019).

3. Methodology

This chapter is divided into two main sections: model description and experimental set-up. The first section addresses the reference biogeochemical model as well as describing the representation of iron in phytoplankton and the various iron parameterisations implemented. The second section will explore the multiple numerical experiments undertaken and the diagnostics utilised for the analysis.

3.1 Model Description

3.1.1 The reference biogeochemical model

Tagliabue *et al.* (2016) only had datasets of existing simulations of various biogeochemical model systems available and the authors could only compare the outputs of these different model formulations and speculate to which degree inter-model differences were due to detail in the iron parameterisations, mainly the scavenging parameterisations. Therefore, the book was constrained and focussed only on biogeochemical models that implemented the parameterisations of Johnson *et al.* (1997) or Parekh *et al.* (2004). To standardise the study, a single biogeochemical model was used and subsequently acted as a testing bench wherein the various parameterisations were embedded. This removed the need to run multiple biogeochemical models and eliminated the issue of inter-model differences in representing other major biogeochemical processes such as carbon uptake or nitrification which have ramifications on the cycling of iron. Therefore, the use of a single model allowed for a more focussed analysis of the various iron parameterisations.

Consequently, the BFM (Vichi *et al.* 2007b, Vichi *et al.* 2015) was chosen to be the main biogeochemical model as its' modular structure allows for the easy inclusion of additional state variables (Vichi *et al.* 2015). The BFM utilises the simpler iron parameterisation of Johnson *et al.* (1997) and employs a hybrid of Monod and quota models in representing the cycling of major nutrients (Vichi *et al.* 2007b), which are present in several current models. In order to analyse the effects of a free scavenging model, like the one described by Parekh *et al.* (2004), a variant of that model employed in a current biogeochemical model was sought. Since a number of biogeochemical models use the scavenging regime of Parekh *et al.* (2004), each with varying alterations and calibrated constants, the iron parameterisations of PISCES

(Aumont *et al.* 2015) were chosen. PISCES uses a free scavenging model and also employs a prognostic appreciation towards ligand concentrations (Tagliabue & Völker 2011). This is an additional facet to describe the scavenging regime that many biogeochemical models do not include yet (Tagliabue *et al.* 2016). In addition to the scavenging scheme, the remineralisation equations of PISCES were also utilised.

BFM description

The BFM stems from the European Regional Seas Ecosystem model (ERSEM) (Baretta, Ebenhöh & Raurdij 1995) and improves upon it by including additional biogeochemical constituents such as iron and chlorophyll which are important components in ocean biogeochemistry (Vichi *et al.* 2007b). The model has been included in several coupled simulations studies (Vichi, Masina & Navarra 2007a; Vichi & Masina 2009; Epicoco *et al.* 2016) which have focused on validating the skill of the BFM. The premise of the BFM is that the functions of producers, decomposers and consumers as well as their respective trophic interactions can be represented in term of material flow of basic elements such as C, N and P.

Taking a functional approach, the BFM defines Chemical Functional Families (CFF) and Living Functional Groups (LFG) which are theoretical constructs used to describe the flow of matter in marine biogeochemistry (Vich *et al.* 2007b). The standard model (Fig. 6) resolves 4 different phytoplankton groups $P^{(j)} = 1,2,3,4$ (diatoms, autotrophic nanoflagellates, picophytoplankton and large phytoplankton), 4 zooplankton $Z^{(j)} = 3,4,5,6$ (carnivorous and omnivorous mesozooplankton, micozooplankton and heterotrophic nanoflagellates), 1 bacteria, 7 inorganic variables for nutrients and gases (phosphate, nitrate, ammonium, silicate, reduction equivalents, oxygen and carbon dioxide) and 10 organic non-living components for dissolved and particulate detritus (Vichi *et al.* 2015).

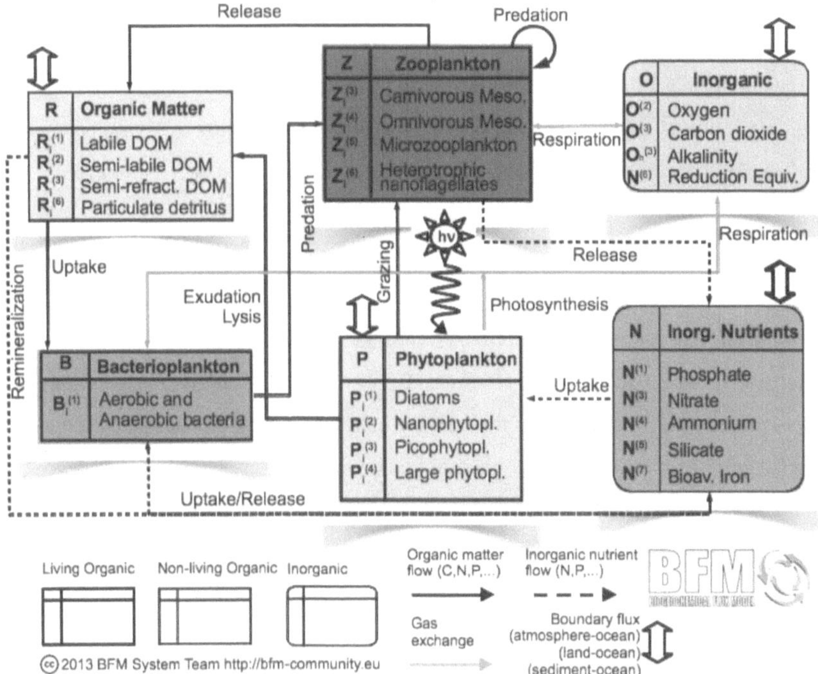

Figure 6: The flow scheme illustrates the interactions of LFGs with organic and inorganic species. CFFs are indicated with bold-line boxes, non-living organic CFFs with thin-line boxes and inorganic CFFs with rounded boxes. (Source: Vichi *et al.* 2015)

Using a theoretical approach, a marine biogeochemical system can be described by the concentrations of CFFs in living and non-living components. Consequently, if C_i indicates a given CFF then the conservation equation for an infinitesimal volume can be written as:

$$\frac{\partial C_i}{\partial t} = -\vec{\nabla} \cdot \vec{F} \tag{2.1}$$

In Eq. (2.1), C_i is continuous in space and time while $\vec{F}$ is the generalised divergence flux of C_i within a fluid. Therefore, Eq. (2.1) can be separated into a physical and biological component:

$$\frac{\partial C_i}{\partial t} = -\nabla \cdot \vec{F}_{phys} - \vec{\nabla} \cdot \vec{F}_{bio} \tag{2.2}$$

The second term on the right hand side of Eq. (2.2) cannot be quantified directly and so the biological component is approximated as follows:

$$\vec{\nabla} \cdot \vec{F}_{bio} = -\omega_B \frac{\partial C_i}{\partial z} + \frac{\partial C_i}{\partial t}\Big|_{bio} \tag{2.3}$$

with the first term parameterising the divergence flux due to sinking of biological particulate matter and the second being the local rate of change of a given C_i by biological processes. Therefore, Eq. (2.3) can be contextualised in the advection-diffusion-reaction equation:

$$\frac{\partial C_i}{\partial t} = \underbrace{-\nabla \cdot (uC_i) + \nabla_H \cdot (A_H \nabla_H C_i)}_{horizontal\ physical\ forcing} + \underbrace{\frac{\partial}{\partial z} A_V \frac{\partial C_i}{\partial z}}_{vertical\ physical\ forcing} - \underbrace{\omega_B \frac{\partial C_i}{\partial z} + \frac{\partial C_i}{\partial t}\Big|_{bio}}_{biological\ forcing} \tag{2.4}$$

Where $u \equiv (u, v, w)$ is the three-dimensional current velocity and (A_H, A_v) are the horizontal and vertical turbulent diffusivity coefficients. Eq. (2.4) is at the basis of biomass based ecosystem modelling where there is the non-local physical forcing of mass by currents and horizontal advection as well as local physical forcing in the form of vertical turbulence. In addition, C_i is altered by various biogeochemical processes.

In the BFM, each variable can be expressed as a multi-dimensional array that contains the concentration of the reference chemical constituents (Vichi *et al.* 2007b). Consequently, a superscript notation indicates the desired CFF for a specific living functional group while a subscript denotes the basic constituent. The example below shows the 6 living CFFs for diatoms:

$$P_i^{(1)} \equiv (P_c^{(1)}, P_n^{(1)}, P_p^{(1)}, P_s^{(1)}, P_l^{(1)}, P_f^{(1)})$$

Following Vichi *et al.* (2007b), the biogeochemical equations represented in Eq. (2.4) can be written in two forms: 1) rates of change; and 2) explicit functional form. For "rates of change" a given CFF state variable C is written as:

$$\frac{\partial C}{\partial t}\Big|_{bio} = \sum_{i=1,n} \sum_{j=1,m} \frac{\partial C}{\partial t}\Big|_{V_i}^{e_j} \tag{2.5}$$

where the right hand side contains the terms representing significant processes for each living and non-living component (Eq. 2.5). The superscript e_j are the abbreviations indicating a specific biogeochemical process (Tab. 1) while the subscript V_i indicates the state variable involved in the process. If a term is present as a source in one equation and a sink in another, the following notation is used:

$$\left.\frac{\partial c}{\partial t}\right|_{V}^{e} = -\left.\frac{\partial V}{\partial t}\right|_{C}^{e} \tag{2.6}$$

Table 1: Showing the various biogeochemical processes in the BFM Source (Vichi *et al.* 2007b)

Abbreviation	**Process**
gpp	Gross primary production
rsp	Respiration
prd	Predation
rel	Biological release: Egestion, Excretion
exu	Exudation
lys	Lysis
syn	Biochemical synthesis
nit/denit	Nitrification, denitrification
scv	Scavenging
rmn	Biochemical remineralisation

In "functional process form" the biogeochemical processes and dependencies are made explicit. Consequently, for ease of reading and understanding, the biogeochemical equations are written in both forms.

3.1.2 Iron dynamics in phytoplankton

Within the BFM, iron is included as an inorganic CFF ($N^{(7)}$), a living organic CFF for phytoplankton as well as a non-living organic CFF for particulate ($R_f^{(6)}$) and dissolved ($R_f^{(1)}$) organic iron (Vichi *et al.* 2015). The iron content of phytoplankton P_f is governed by three primary processes: uptake of bioavailable iron, loss due to lysis as well as predation by zooplankton (Eq. 2.7).

$$\frac{\partial P_f}{\partial t} = \left.\frac{\partial P_f}{\partial t}\right|_{N^{(7)}}^{upt} - \left.\frac{\partial P_f}{\partial t}\right|_{R_f^{(6)}}^{lys} - \frac{P_f}{P_c}\sum_{k=3}^{6}\left.\frac{\partial P_c}{\partial t}\right|_{Z_c^{(k)}}^{prd} \tag{2.7}$$

The iron uptake shown in Eq. (2.8) is computed according to Droop kinetics by taking the minimum of two rates, a linear function of the ambient concentration representing the membrane through-flow at low concentrations and a balancing flux linked to carbon assimilation. The second term pertains more to coastal systems/upwelling sites where high macronutrient concentrations occur; facilitating high biological productivity. Consequently, the iron:carbon quota for phytoplankton varies between a maximum (ϕ_P^{max}) and minimum (ϕ_P^{min}) threshold where ϕ_P^{min} represents the adaptive ability of each functional group at prevailing iron concentrations and ϕ_P^{opt} is the optimal ratio for growth.

$$\frac{\partial P_f}{\partial t}\Big|_{N^{(7)}}^{upt} = min(a_p^7 N^{(7)} P_c, \phi_p^{opt} G_p + f_p^T r_p^0 (\phi_p^{max} - \frac{P_f}{P_c}) P_c) \qquad (2.8)$$

Meanwhile, it is assumed that the only physiological iron loss from phytoplankton is due to cell disruption and that when a cell is about to die, it has the minimum quota of iron ϕ_P^{min} present. Consequently, iron loss is computed according to carbon lysis multiplied by ϕ_P^{min} as shown in Eq. (2.9).

$$\frac{\partial P_f}{\partial t}\Big|_{R_f^{(6)}}^{lys} = \phi_P^{min} \frac{\partial P_c}{\partial t}\Big|_{R_c^{(6)}}^{lys} \qquad (2.9)$$

Therefore, the phytoplankton processes of uptake and cell lysis form an important part in utilising $N^{(7)}$ and controlling the flux of $R_f^{(6)}$ and $R_f^{(1)}$.

With regards to the work of Wagner *et al.* (2008), refer to Eq. (2.17), the production of DOC is an important parameterisation for determining the concentration of organic ligands. Consequently, the production pathway of DOC ($R_c^{(1)}$) is governed by three primary processes: exudation of carbohydrates from phytoplankton, uptake by bacteria as well as release of $R_c^{(1)}$ by zooplankton (Eq. 2.10).

$$\frac{\partial R_c^{(1)}}{\partial t}\Big|_{bio} = \sum_{j=1}^{3} \frac{\partial P_c^{(j)}}{\partial t}\Big|_{R_1^{(1)}}^{exu} - \frac{\partial B_c}{\partial t}\Big|_{R_c^{(1)}}^{upt} + \sum_{k=5,6} \frac{\partial Z_c^{(k)}}{\partial t}\Big|_{R_c^{(1)}}^{rel} \qquad (2.10)$$

Focussing on the first term of Eq. (2.10), it is assumed that when there are intra-cellular nutrient shortages, not all photosynthesised carbon can be assimilated. Consequently, the non-assimilated portion is released as DOC. Therefore, increased exudation is observed under nutrient-stress conditions when the nutrient:carbon ratio becomes low (Vichi *et al.* 2007b).

3.1.3 Iron parameterisations

Similar to BFM, PISCES simulates marine biological productivity and describes the biogeochemical cyclings of major nutrients (P, N, Si, Fe). There are 24 prognostic variables in PISCES; encompassing phytoplankton, zooplankton, inorganic and organic nutrients (refer to Aumont *et al.* 2015). As well as representing the major processes of the iron cycle such as scavenging, remineralisation and uptake by phytoplankton, PISCES incorporates several additional processes that are not present in the BFM such as: uptake of iron by bacteria, colloidal fractions and aggregation (Aumont *et al.* 2015). In addition, PISCES has two chemistry models for iron: a simple model based on one ligand and one inorganic iron species and a complex model that uses five iron species and two ligand classes. To amalgamate the PISCES parameterisations into BFM, the simple model was used for ease of translation.

To compare the iron formalisms of scavenging and remineralisation between the two models, the PISCES equations, as shown in Aumont *et al.* (2015), were written in an explicit functional form. For consistency, the PISCES equations were translated into the format of the BFM, substituting the PISCES variables for the appropriate BFM variable and only adding additional variables and diagnostics where necessary. Furthermore, the parameter values for the scavenging and remineralisation rates for PISCES were conserved when translating the model formalisms.

BFM iron dynamics

The concentration of bioavailable iron for the BFM, shown in Eq. (2.11), is influenced by: the uptake of iron by phytoplankton, the remineralisation of particulate and labile dissolved organic material as well as the scavenging of dissolved iron. Consequently, Eq. (2.11) forms the basis of the iron cycling model within BFM.

$$\frac{\partial N^{(7)}}{\partial t} = -\frac{\partial P_f}{\partial t}\Big|_{N^{(7)}}^{upt} + \sum\nolimits_{i=1,6}\frac{\partial R_f^{(i)}}{\partial t}\Big|_{N^{(7)}}^{rmn} - \frac{\partial N^{(7)}}{\partial t}\Big|_{sinkf}^{scv} \tag{2.11}$$

Remineralisation

In a marine ecosystem, the remineralisation of particulate and dissolved organic matter is facilitated by autotrophic and heterotrophic bacteria. The remineralisation scheme for the BFM is a linear function for both particulate and labile (dissolved) organic iron (Eq. 2.12).

$$\left.\frac{\partial R_f^{(i)}}{\partial t}\right|_{N^{(7)}}^{rmn}{}_{i=1,6} = \Lambda_f^{rmn} f_{R_f^{(6)}}^{T} R_f^{(i)} \tag{2.12}$$

There is a temperature dependence term $f_{R_f^{(6)}}^{T}$ and a constant remineralisation rate Λ_f^{rmn}, where the remineralisation rate of dissolved iron is an order of magnitude greater than the particulate. However, a drawback is that the linear formulation decouples iron from other major nutrients such as N and P, which are instead dynamically remineralised by bacteria.

Consequently, Eq. (2.13) shows that the concentration of non-living organic iron species in the BFM are influenced by the lysis of phytoplankton and coupled to this, the predation of phytoplankton by zooplankton as well as the remineralisation of non-living organic iron species (Eq. 2.12). It is important to note that iron is not tracked within zooplankton and it is assumed that zooplankton is never iron-limited, with the iron fraction of ingested phytoplankton being egested as particulate detritus (Vichi *et al.* 2007b).

$$\left.\frac{\partial R_f^{(i)}}{\partial t}\right|_{bio}{}_{i=1,6} = \left.\frac{\partial P_f}{\partial t}\right|_{R_f^{(i)}}^{lys} + \frac{P_f}{P_c}\sum_{k=3}^{6}\left.\frac{\partial P_c}{\partial t}\right|_{Z_c^{(k)}}^{prd} - \sum_{i=1,6}\left.\frac{\partial R_f^{(i)}}{\partial t}\right|_{N^{(7)}}^{rmn} \tag{2.13}$$

Scavenging

The scavenging dynamics for BFM consider inorganic and organic mechanisms as well as the buffering effect of ligands (Eq. 2.14). Consequently, the formalism follows that of Johnson *et al.* (1997), assuming a single strong iron binding ligand that controls the solubility of iron when $N^{(7)}$ exceeds 0.6 nM. In addition, the BFM also considers the scavenging effect of sinking detrital matter, represented by particulate organic carbon ($R_c^{(6)}$), that acts to absorb dissolved iron and transport it into the deep ocean.

Two scavenging constants are used: Λ_f^{scvorg} for scavenging and absorption onto particles and Λ_f^{scv} for the buffering effect of ligands. Under the current formulation, scavenging is always larger than zero, permitting scavenging at any concentration of dissolved inorganic iron even if there is no particulate detritus.

$$\left.\frac{\partial N^{(7)}}{\partial t}\right|_{sinkf}^{scv} = max(0, \Lambda_f^{scvorg} N^{(7)} {R_c^{(6)}}^{0.58}) + \Lambda_f^{scv} max(0, N^{(7)} - 0.6) \quad (2.14)$$

<u>*PISCES iron dynamics*</u>

The iron parameterisations of PISCES were translated and added as a separate module within the BFM. The objective in creating the PISCES iron module was to conserve as much of the BFM source code as possible while incorporating as many elements of the PISCES iron dynamics. However, PISCES included several processes not present in the BFM such as colloidal interactions, which were not added, but are important as they affect both free and complexed species of iron and can be a significant abiotic loss term (Aumont *et al.* 2015).

Remineralisation

The PISCES formalism for the remineralisation of particulate organic iron (Eq. 2.15a) is similar to that of BFM's (Eq. 2.12) as both schemes are simple linear functions.

$$\left.\frac{\partial R_f^{(6)}}{\partial t}\right|_{N^{(7)}}^{rmn} = \Lambda_f^{rmn} f_{R_f^{(6)}}^{T} (1 - 0.45\Delta(O^{(2)})) R_f^{(6)} \quad (2.15a)$$

$$\Delta(O_2) = min(1, max(0, 0.4 \frac{O_2^{min,1} - O^{(2)}}{O_2^{min,2} + O^{(2)}})) \quad (2.15b)$$

An additional facet to the remineralisation scheme for PISCES is an environmental oxygen dependency term $\Delta(O^{(2)})$. $\Delta(O^{(2)})$ was added as a diagnostic into the BFM (Eq. 2.15b) and it can vary between 0 (oxic) and 1 (anoxia). During oxic conditions, the rate of remineralisation would be greater for a given temperature. To account for the fact that PISCES does not include a tracer variable for dissolved organic iron, the BFM scheme of $R_f^{(1)}$ was maintained.

Scavenging

The simple PISCES iron chemistry model (Aumont *et al.* 2015) uses one ligand class and two dissolved iron species: dissolved inorganic iron and dissolved complexed iron, in accordance with the free scavenging model of Parekh *et al.* (2004). Both forms of iron are susceptible to consumption by phytoplankton and the total bioavailable iron concentration is the sum of the non-complexed and complexed dissolved iron species. To represent the free scavenging model in the BFM, an additional diagnostic was created, $N_{free}^{(7)}$ and only this species of iron would be susceptible to scavenging.

$$\Delta = 1 + K_{N^{(7)}}^{eq} L_T - K_{N^{(7)}}^{eq} N^{(7)}$$

$$N_{free}^{(7)} = \frac{-\Delta + \sqrt{(\Delta^2 + 4K_{N^{(7)}}^{eq} N^{(7)})}}{2K_{N^{(7)}}^{eq}} \tag{2.16}$$

The concentration of $N_{free}^{(7)}$ (Eq. 2.16) is computed using a chemical equilibrium constant for iron in seawater ($K_{N^{(7)}}^{eq}$ adopted from PISCES) as well the total ligand concentration L_T.

Unlike the BFM that assumes a constant ligand concentration, PISCES allows for a prognostic (Eq. 2.17) as well as an explicit concentration (Aumont *et al.* 2015) using the relationship from Tagliabue & Völker (2015). Eq. (2.17) is built upon the work of Wagener *et al.* (2008) who showed a relationship between DOC and ligand concentrations. Consequently, a switch function is employed to ensure that even under low biological activity, L_T will be at least 0.6 nM.

$$L_T = max(0.09(R_c^{(1)} + 40) - 3,\ 0.6) \tag{2.17}$$

Similar to BFM, PISCES uses organic and inorganic mechanisms in the scavenging regime as shown in Eq. (2.18).

$$\Lambda_f^{scvtot} = \Lambda_f^{scvmin} + \Lambda_f^{scvorg}(R_c^{(6)} + O_c^{(5)} + R_s^{(6)}) + \Lambda_f^{dust} Dust \tag{2.18}$$

$$\frac{\partial N^{(7)}}{\partial t}\Big|_{sinkf}^{scv} = \Lambda_f^{scvtot} N_{free}^{(7)} \tag{2.19}$$

Unlike the BFM, the scavenging rate constant is treated as a variable in PISCES (Λ_f^{scvtot}) and consists of a minimum scavenging rate (Λ_f^{scvmin}) as well as the total particulate load of the seawater which is separated into biogenic and lithogenic particles. This approach of including biogenic and lithogenic particles is similar to that of Moore & Braucher (2008) in the BEC model. It is assumed that the scavenging rates of biogenic (Λ_f^{scvorg}) and lithogenic (Λ_f^{dust}) particles are different because they have dissimilar affinities for iron.

The amount of iron scavenged is heavily influenced by biological activity which will determine the total ligand concentration as well as the biogenic particulate load of the water. Consequently, PISCES employs a more dynamic range of interactions than the BFM, facilitating a wider range of modelling capabilities.

3.2 Experimental set-up

3.2.1 Model set-up

The diverse physical, chemical and biological conditions in the oceans lend themselves to harbouring unique biogeochemical systems. Consequently, four disparate regions were selected for study: North Atlantic gyre (NAG), SO, EP and the NEP under typical mixed layer conditions. The oligotrophic NAG is of interest as it experiences some of the highest rates of dust deposition (Fig. 7) in the global ocean (Jickells *et al.* 2005; Anderson *et al.* 2016) as well as being a region where phosphorous is a limiting nutrient for N_2 fixation (Mills *et al.* 2004). Whereas the SO (Martin *et al.* 1991), NEP (Martin & Fitzwater 1988) and EP (Fitzwater *et al.* 1996) are the major HNLC regions (Fig. 7) and are iron deficient due to low rates of dust deposition (Jickells *et al.* 2005) which is a limiting factor for primary productivity. Although the SO, NEP and EP are similar in terms of excess major nutrients, they differ in their physical oceanographic properties of Sea-Surface Temperature (SST), light availability, salinity and Mixed-Layer Depth (MLD) which will influence their respective biogeochemical processes.

Modifications to standard BFM

In order to assess the response of the two iron parameterisations, the BFM was configured to run in a standalone 0D/box-model format. Fennel & Neumann (2001) highlighted that simple box-models allow for the identification of key processes in a biogeochemical system which was exactly of interest. In addition, biogeochemical parametrisations are independent of spatial resolution. However, a major shortfall in using a 0D configuration was the inability to represent the key physical processes of horizontal and vertical advection of nutrients as well as surface-boundary layer fluxes such as dust deposition. Consequently, the standard forcing functions and boundary conditions of the BFM model as described by Vichi *et al.* (2015) were extended. Three additional components were added to the model: a variable MLD as well as a boundary condition for dust deposition and a dust particle state variable. This was done with the intention of facilitating the inclusion of the PISCES formalisms into BFM as well as to add some important physical dynamics to the 0D simulations.

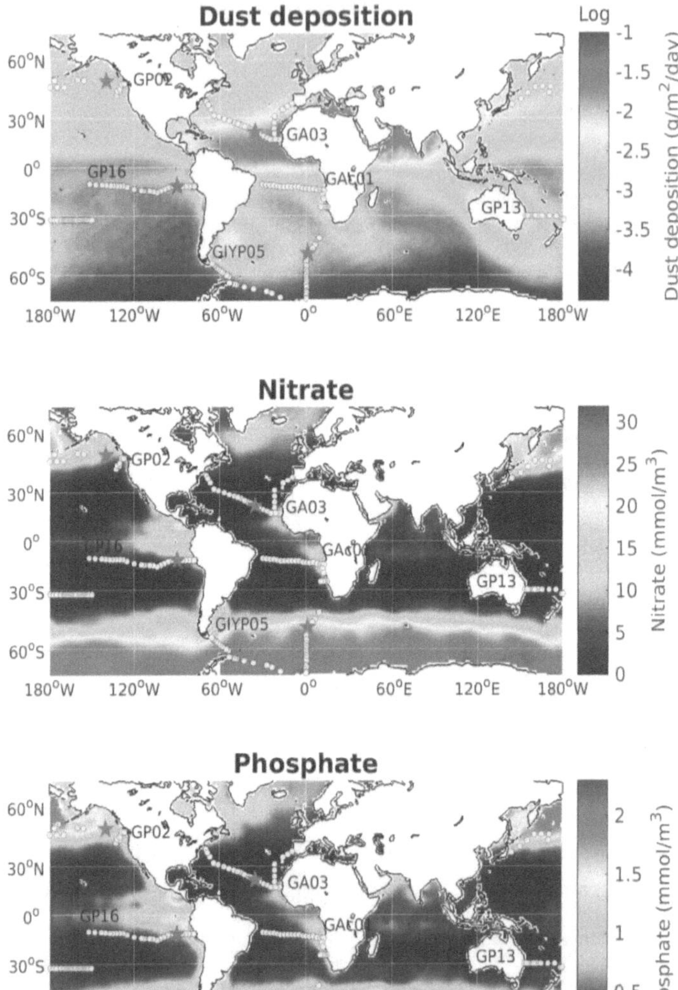

Figure 7: The annual surface mean values for: dust deposition rate (Jickells et al. 2005), nitrate (WOA18) and phosphate (WOA18) concentrations. Several cruises from the GEOTRACES programme (Schlitzer et al. 2018) are shown; with a grey dot representing a location where a full depth CTD iron profile was taken. Red stars indicate the location where the BFM was run.

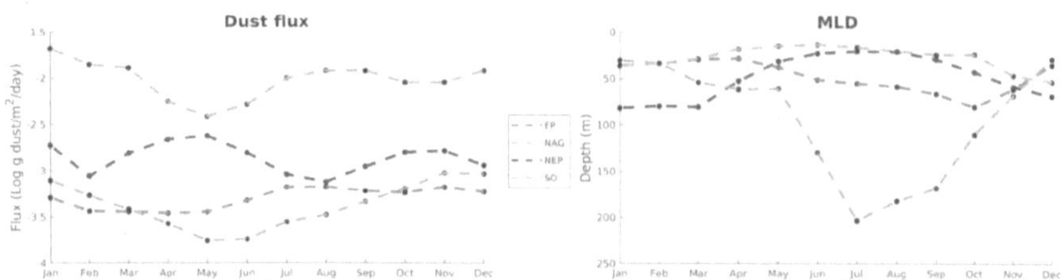

Figure 8: Climatology of surface dust deposition (left) (Jickells et al. 2005) and MLD (right) for the NAG, EP, SO (de Boyer Montégut et al. 2004) and NEP (Holte et al. 2017).

Dust deposition model

In selecting the regions for the study (Fig. 7), only open ocean locations were chosen as coastal regions have additional sources of iron such as riverine and sediment deposits (Boyd & Ellwood 2010) which are difficult to constrain in a 0D simulation. Consequently, for the various regions, atmospheric dust was the principle source of iron. For each location the surface-boundary flux for mineral dust (Fig. 8) was taken from the climatological model of Jickells *et al.* (2005) where it was assumed that the average iron content of mineral dust was 2% with a dissolution fraction of 3.5%. To complement the surface-boundary flux of atmospheric dust, a state variable for dust was added to the BFM as it plays a role in the scavenging of dissolved iron. The concentration of $Dust$ was computed using the formalism (Eq. 2.20) of Aumont *et al.* (2015) which used the surface deposition D_{dust} and sinking speed ω_{dust} of dust. In a 0D model, this assumes a homogeneous distribution of dust within the box, independently of its depth.

$$Dust = \frac{D_{dust}}{\omega_{dust}} \tag{2.20}$$

Variable MLD

The standalone configuration for the BFM focussed on representing the biogeochemical dynamics for the upper-ocean which was constrained to the depth of the MLD. By using a box-model approach, it is assumed that the processes of uptake, remineralisation and scavenging are confined to the upper-ocean and thus the biogeochemical system is sustained through recycled nutrients. However, for the upper-ocean, vertical mixing and upwelling act as important sources of new nutrients (Falkowski *et al.* 1998; Fung *et al.* 2000). The inclusion of additional sources of preformed iron would however influence the rigorous investigation of the parameterisations. Acknowledging the fact that the shoaling and deepening of the MLD also plays an important role in influencing the nutrient concentrations for the euphotic zone, the addition of a variable MLD in the BFM was done to take into consideration the changes in light availability to phytoplankton and the length scale of sinking particles. Rudimental methods to scale the input of organic nutrients based on the gradient of the MLD would result in the addition of spurious modes of variability which would be counter-productive in efforts to identify the main processes that control the iron dynamics. Therefore, the inability to represent the physical dynamics of upwelling of nutrients would be a major shortfall when modelling the EP as the region is heavily influenced by episodic upwelling events which are driven by the easterly trade winds (Bidigare & Ondrusek 1996).

For each location (Fig. 8), the MLD data was extracted from the monthly climatology data of de Boyer Montégut et al. (2004) (available at www.ifremer.fr) for the NAG, EP and SO while data for the NEP was sourced from Holte *et al.* (2017) (available at mixedlayer.ucsd). The main reason the NEP used a different data set to the other regions was due to the MLD climatology of de Boyer Montégut *et al.* (2004) poorly resolving the depth and temporal extent of the Winter time MLD for the region while Holte *et al.* (2017) produced more consistent results with other observational works, such as with Ohno *et al.* (2009).

3.2.2 Model analysis

In the following, the use of the term "FeBFM" will refer to the BFM model with the iron dynamics of Vichi *et al.* (2015) while "FePISCES" to the iron parameterisations of Aumont *et al.* (2015) that have been adjusted to be compatible with the BFM source code. Once the PISCES iron parameterisations were embedded into the BFM; a total of eight simulations were run. Each simulation was initialised in January and ran for 10 years to allow the biogeochemical system to reach a steady state. FeBFM and FePISCES were each run once in four locations, representative of the: NAG, SO, NEP and EP (Fig. 7).

Model initialisation

For each configuration, FeBFM and FePISCES, the initial conditions for the macronutrient concentrations and physical forcing were identical. The World Ocean Atlas 2018 (WOA18) was used for the macronutrient concentrations of: oxygen, (Garcia *et al.* 2018a) nitrate, phosphate and silicate (Garcia *et al.* 2018b) while the physical forcing conditions of light intensity, wind speed, SST and salinity are shown in Tab. (2). For the WOA18 and physical forcing data sets, the Climate Data Operators (Schulzweida 2019) (CDO) program was used to create the forcing files for the simulations.

Table 2: List of data sets used for the physical forcing conditions in the BFM

Variable	Data set	Reference
SST	NOAA_OI_SST_V2	www.esrl.noaa.gov
Light	NCEP/DOE 2 Reanalysis data	www.esrl.noaa.gov
Wind	NCEP/DOE 2 Reanalysis data	www.esrl.noaa.gov
Salinity	World Ocean Atlas 2013	www.nodc.noaa.gov

The iron data was sourced from the GEOTRACES IDP2017v1 (Schlitzer *et al.* 2018) with the NAG, SO, NEP and EP corresponding to the: GA03, GIPY05, GP02 and GP16 GEOTRACES' cruises (Fig. 7). Owing to the scarcity of iron data, from each location, surface iron data was extracted from stations within a 400 km radius, corresponding to the resolution of 2° earth system models (McKiver *et al.* 2015). Each configuration was initialised with homogeneous initial conditions for all the LFGs and the BFM state variables were the same ones as used in the global simulations by Vichi *et al.* (2007a, b).

Metrics utilised

A wide variety of methods can be employed to analyse model outputs; however, choosing the best metrics as well as understanding the score itself is not trivial (Stow *et al.* 2009). Taylor diagrams (Taylor 2001) are ideal for juxtaposing observational data with model outputs; however, in comparing and contrasting the model runs of FeBFM and FePISCES, the objective was not to validate either configuration but to assess the representation of dissolved iron as well as the cascading effect on the wider biogeochemical community. Consequently, two methods were used: time series plots and Principal Component Analysis (PCA).

Time series

Stow *et al.* (2009) noted that a major feature in comparing ecosystem models was to identify the appearance of specific features and/or patterns in the model outputs and observational data. Consequently, the model outputs of FeBFM and FePISCES were compared against climatology data from the WOA18 and Ocean Colour Data from the European Space Agency (ESA) (available at esa cci). The objective in using the time series plots was to assess whether altering the iron parameterisations would affect the seasonal cycling of nutrients. However, no climatology data existed for dissolved iron, so only the macronutrients of: nitrate, phosphate and silicate were studied as well as chlorophyll.

PCA plots

In 1901, Pearson developed PCA as an explanatory technique aimed at identifying unknown trends in multidimensional data sets (Abdi & Williams 2010). PCA utilises singular value decomposition from linear algebra to decompose a square correlation matrix. The left and right eigenvectors as well as singular values allow for the relationship between two variables in multidimensional space to be assessed using a smaller number of principal components. The power of PCA is in its ability to reduce the number of dimensions in a data set. Consequently, for multi-variable systems, PCA allows for the relationship between variables to be determined by their relative position in a lower dimension space (usually 2D). This allows for an easy assessment of how the different iron parameterisations affect the whole biogeochemical system without relying on multiple time series plots. However, unlike the time series plots, PCA was reserved for the model outputs as insufficient observational data limited its' application. Therefore, the application was aimed at conducting an inter-comparison between FeBFM and FePISCES.

When conducting a PCA, the correlation matrix requires that data be sampled from a normal distribution. Inherently, natural systems and model outputs struggle to be Gaussian owing to the high number of non-linear interactions. Therefore, non-Gaussian distributed data must be transformed before applying PCA. On analysing the distribution types for the various state variables in the model outputs, the most observed were: normal, log-normal, generalised Pareto and extreme value. Subsequently, applying a transformation such as the box-cox to all the non-Gaussian state variables seemed excessive. Instead, noting that most of the variables were positively skewed, it was assumed that the dominant distribution was log-normal and thus taking the natural logarithm of the model output data would transform the state variables to a normal distribution.

To illustrate how to read and understand a PCA plot, an example is shown in Fig. (9), but this does not intend to be an exhaustive explanation about PCA and instead aims to showcases the basics that will become important in Sec. (4.2.3).

The example plot shows various characteristics pertinent to cars. The data matrix is decomposed using two principal components (PC1-2) with the horizontal axis showing the projections to PC1 while the vertical axis is PC2. The percentage explained for each PC is an indication of the percentage of the total variance each PC explains. Therefore, from Fig. (8), PC1 explains 62.8% of the total variance while PC2 explains 23.1%. The variables are plotted as vectors from the origin and their orientation is an indication of their influence by a specific PC. For example, the number of gears in a car is strongly influenced by PC2 while miles per gallon (mpg) is influenced by PC1. In addition, the cosine of the angles between the vectors is an indication of the correlation between each variable. Therefore, there is a strong positive correlation between the number of cylinders a car has (cyl) and the combined volume of an engine's cylinders (disp), while there is a negative correlation between the number of cylinders and the mpg of a car and no correlation between the number of gears and number of carburettors (carb) a car has. The black dots are the scores which in this example correspond to the various car models. The proximity of a score relative to the head of a vector indicates how much variance a single vector describes. A good example is for mpg, where there are three scores in close proximity. This means that those three cars share similar characteristics and are strongly influenced by mpg.

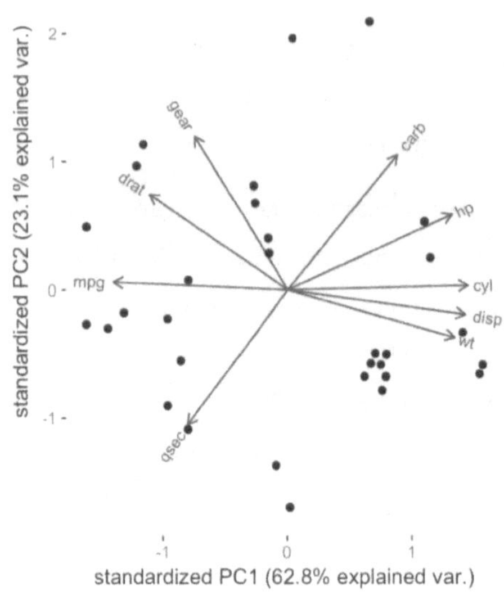

Figure 9: Example PCA plot showing characteristics of various cars (Source: www.datacamp.com)

4. Results

The results chapter is divided into two main sections: iron cycling of FeBFM vs FePISCES and biogeochemical impact. The first section compares and contrasts the cycling of iron between the two formulations described in Sec. (3.1.3). The objective is to highlight the implications of the different scavenging parameterisations between the two configurations. The second section then builds upon the first by exploring the impacts of the different iron configurations on the model states, such as: community composition, macronutrient cycling as well as state variable relationships as described in Sec (3.2.2).

4.1 Iron cycling of FeBFM vs FePISCES

The biogeochemical cycling of iron involves multiple abiotic and biotic processes such as scavenging, remineralisation, lithogenic inputs and biological uptake (see Sec. 2.1.1) and these processes need to be modelled and parameterised in biogeochemical models. To assess the broad impacts of the iron parameterisations on the dynamics of iron between the two configurations, the residence times and annual mean concentrations of dissolved iron as well as the annual mean flux of dust for the various modelled regions are shown in Tab. (3). The residence times were calculated by dividing the depth integrated mean iron concentration over ten years by the mean flux of atmospheric dust over the same period. Consequently, the residence times provided an important metric in gauging how rapidly iron was cycled in the various locations, which would be a reflection upon the behaviour of the different iron parameterisations. Short residence times are associated with high dust deposition regions with high scavenging rates while long residence times occur in low deposition regions.

Table 3: Concentration and residence time of iron in the modelled regions

	FeBFM			FePISCES	
	Residence time (years)	Mean iron (nM)	**Mean dust flux ($g/m^2/year$)**	Residence time (years)	Mean iron (nM)
SO	10.37	0.17	**0.15**	21.27	0.33
EP	4.0	0.19	**0.21**	10.08	0.47
NAG	0.72	1.07	**3.87**	1.12	1.65
NEP	3.08	0.34	**0.52**	4.88	0.55

For all the modelled regions, FePISCES observed longer residence times as well as dissolved iron concentrations that were two to three times greater than FeBFM's. As both configurations had an identical dust flux, only the differences in their scavenging and remineralisation parameterisations (Sec. 4.1.2) were responsible for affecting the concentration and cycling of iron. Both configurations showed low dissolved iron concentrations occurring in the HNLC regions of the SO, EP and NEP and elevated concentrations in the high dust deposition region of the NAG. As expected, the low dust deposition regions of the SO and EP had longer residence times compared to the NEP and NAG, with both configurations having residence times that were comparable to the work of Moore *et al.* (2004).

4.1.1 Iron time-series analysis

To appreciate how the different iron parameterisations affected the concentration of iron, it was necessary to visualise the seasonal cycle as well as the long-term evolution of iron in the model runs. Fig. (10) compares and contrasts the seasonal and ten-year cycling of iron between FePISCES and FeBFM in the modelled regions. Starting with the climatologies, FePISCES had a greater seasonal variability of iron than FeBFM in all the model locations. However, both configurations showed similar seasonal cycles for iron with both capturing the maxima of iron in September for the NAG as well as the relatively constant iron concentrations of the EP. In the SO, the seasonal variability was greater in FePISCES, but both configurations showed a summertime minimum (December-January) and a winter maximum (June-August), following the trend of biological productivity for the region. For the NEP, both configurations showed a minimum iron concentration in March and a maximum in August which was anomalous because the NEP is in the northern hemisphere and it was expected that the minimum concentration of iron would occur in September, corresponding to the summer months and a maximum in winter, around January, due to accumulation and remineralisation.

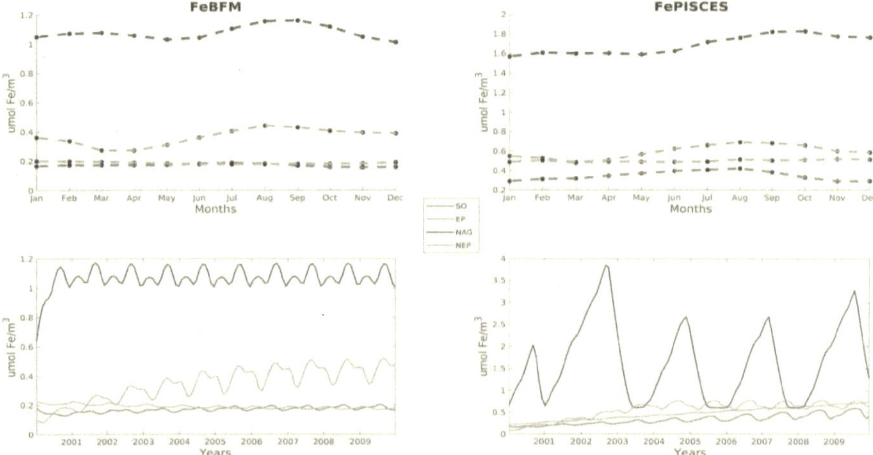

Figure 10: (upper row) Annual climatology of iron. (lower row) Ten-year time-series of iron

To further understand the cycling dynamics for each region, it was necessary to isolate the processes responsible for controlling the seasonal variability. Of importance was the influence of the dust flux in controlling the seasonal cycle of iron. Tab. (3) shows the mean annual flux of dust and it was expected that in regions of high atmospheric deposition, there would be a good correspondence between dissolved iron concentrations and lithogenic inputs. Conducting a Pearson's correlation between iron and the dust flux revealed a surprising result. For the SO, both FeBFM and FePISCES had strong negative correlations (-0.71 and -0.60) while the highest positive correlation occurred in the EP (0.55 and 0.43). Meanwhile, the NAG had a moderate correlation of 0.45 with FeBFM but a significantly lower one of 0.26 in FePISCES. In the NEP, the correlations were the smallest of any regions at 0.18 for FeBFM and 0.16 for FePISCES.

Therefore, in the low to moderate correlation regions of the EP, NAG and NEP, the deposition of dust was not playing a significant role in driving the cycling of iron. However, in the SO, the strong negative correlation could be significant as the onset of high dust deposition in the summer months could seed elevated levels of biological activity and hence result in the consumption of available iron. Consequently, the assumption that high dust deposition regions would have elevated correlations with iron was unfounded, as this was exemplified in the NEP as well as the NAG for the FePISCES configuration. Thus, the cycling of iron for the respective regions must be driven by the chosen parameterisations (scavenging and remineralisation dynamics of the respective configurations).

While appreciating the seasonal cycling of iron, it was necessary to understand the long-term variability of iron in the box-model simulations, because this set-up is a coarse approximation of what may happen in reality and in 3D coupled models. Similar to the climatology plots, FePISCES observed greater variability than FeBFM, especially in the NAG. A key difference between FeBFM and FePISCES was that FeBFM reached a steady-state in all the modelled regions except in the NEP, while FePISCES did not reach a state of equilibrium in any location. Fig. (10) shows that in all the modelled regions, FePISCES had a steady upward gradient for the SO, EP and NEP while the NAG displayed a biennial oscillation. The biennial oscillation in the NAG is likely to stem from a mathematical feedback in the system of PDEs since it isn't forced by any external iron. However, the lack of a steady state may have been responsible for the lower correlations with atmospheric dust in comparison with FeBFM. It should also be noted that spurious trends and long-term cycles may create local artificial gradients in a coupled configuration with a transport model. Therefore, the long-term variability of iron was influenced by the iron configuration that was used. Since all the inputs and initial conditions were the same, variability in the iron cycles between FeBFM and

FePISCES can be attributed to the respective scavenging and remineralisation parameterisations. Consequently, the next section explores the role of the scavenging and remineralisation processes and how they affect the seasonality and cycling of iron.

4.1.2 Scavenging and remineralisation dynamics

The contrast in the time-series plots for the two configurations (Fig. 10) was attributed to the differences in the respective remineralisation and scavenging parameterisations. Thus, Fig. (11) compares the scavenging and remineralisation rates for FeBFM and FePISCES in the modelled regions to ascertain what implications the parameterisations had on controlling the cycling of iron. In reality, the true scavenging and remineralisation rates are unknown and therefore this is a theoretical exercise to understand the functions.

The HNLC regions, especially the SO and NEP, had higher remineralisation rates than the NAG due to elevated levels of biological activity and nutrient availability. Whereas the scavenging rate was proportional to the concentration of dissolved iron and thus the NAG had the greatest scavenging rate compared to the SO and EP. Referring back to Sec. (3.1.3), the main difference between FeBFM and FePISCES stemmed from their different formulations regarding the scavenging and remineralisation dynamics of iron. Consequently, Fig. (11) highlights the difference in behaviour between the two configurations.

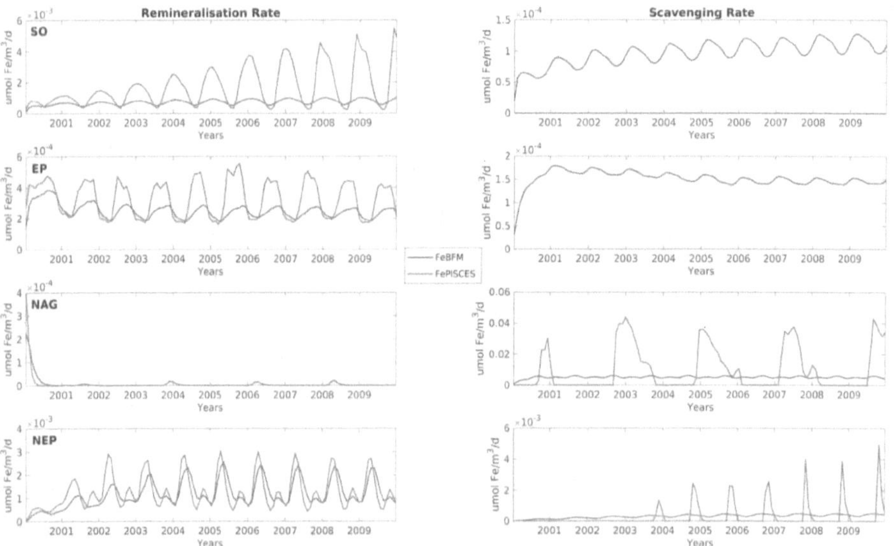

Figure 11: The remineralisation (left) and scavenging (right) rates for FeBFM (blue) and FePISCES (red)

Starting in the SO, both configurations observed higher remineralisation rates in summer and lower in winter; however, FeBFM maintained a lower but steady-state remineralisation rate in comparison to FePISCES; which grew in magnitude with time. As for the scavenging rate, FePISCES had little to no scavenging throughout the entire simulation while FeBFM observed a scavenging trend consistent with the seasonality of iron. Consequently, the lack of scavenging resulted in the accumulation of iron in the SO for FePISCES. As for the remineralisation rate, the steady increase in FePISCES reflected an increase in the production of particulate matter which could only be induced by increasing levels of primary productivity as the simulation evolved (See Fig. 13). The EP was similar to the SO as FePISCES maintained a greater remineralisation rate than FeBFM but in contrast to the SO, FePISCES did not have an increasing rate as the simulation evolved. Instead, the EP saw a convex shape regarding the evolution of the maxima for FePISCES. Again, FePISCES had little to no scavenging throughout the simulation which caused iron to accumulate while FeBFM maintained a relatively steady-state at lower ambient iron concentrations.

The NAG presented more of a challenge when interpreting the results as the oligotrophic state of the region resulted in very little primary productivity in the model simulations (refer to Fig. 12 and 13). For the remineralisation regime, both configurations had near identical time-series, very close to zero. The lack of biological activity due to limited nutrient availability inherently resulted in little particulate matter production and thus low remineralisation rates. However, a large contrast was observed in the scavenging regime linked to the substantial dust flux in the region (Tab. 3). The biennial oscillation of scavenging for FePISCES is out of phase with the seasonal cycle of iron for the region (Fig. 10). Consequently, when dissolved iron reaches a threshold concentration, scavenging is activated and reduces the dissolved iron concentration. Despite the repeating seasonal cycle of dust deposition (Fig. 8), which does not have any biennial oscillation, the scavenging rate seems to have a threshold response that triggers the cycle. As for FeBFM, the scavenging rate remained relatively constant throughout the simulation.

The NEP saw FeBFM and FePISCES following a very similar remineralisation time-series. However, two features stand-out, the amplitude of FePISCES was greater than FeBFM and the phasing of the two configurations were not directly in sync. The relatively steady-state of the remineralisation regime for FePISCES was in contrast to the behaviour of the EP and SO which had underlying trends. Focussing on the scavenging regime, for FePISCES in the NEP, the fluctuation was similar to the NAG, where there were periods of intense scavenging followed by a lull period of little to no scavenging. While FeBFM maintained a lower steady-state scavenging rate compared to FePISCES, FePISCES was punctuated by annual events of intensive scavenging. Similar to the NAG, the onset of scavenging occurred when dissolved iron concentrations reached a threshold value.

It was apparent that the remineralisation rates for FeBFM and FePISCES were more similar than that of the scavenging. However, the scavenging regimes of FePISCES and FeBFM differ significantly in-terms of their representation of organic ligands as well as the scavenging model employed. Therefore, it was apparent that the choice of scavenging dynamics had distinct behaviours and these inherently altered the cycling of iron in the modelled regions. However, what implications would different iron cycling have on the general behaviour of the biological community?

4.2 Biogeochemical impact

To understand the implications of the iron cycling on the other biogeochemical state variables within the BFM, this section is divided into four parts. The first will analyse phytoplankton biomasses and assemblages while the second section will explore the differences in the cycling of macronutrients. The third section will attempt to identify what impact the iron parameterisations have on the state variable relationships within the BFM. Finally, the last section will investigate the effect of ligands in driving the cycling dynamics of iron (refer to Sec. 3.1.3).

4.2.1 Impact on phytoplankton community composition

Fig. (12) summarises the differences between FeBFM and FePISCES in terms of phytoplankton community composition and cumulative biological productivity for the four selected regions over the ten-year model simulation.

The HNLC regions had the most biological activity, with the SO having the highest biomass of any location. Between FeBFM and FePISCES, there was not a considerable difference in the total biomass in each region, considering that iron concentrations were nearly double in FePISCES compared to FeBFM (Tab. 3). These observations suggest that iron may not have been the limiting nutrient for growth in some regions, thus additional nutrient limitations may have been responsible for hindering significant biological activity. In the SO, FePISCES observed a greater abundance of diatoms than FeBFM; however, this was accompanied by smaller picophytoplanton and flagellate communities. In the EP, FeBFM had a slightly greater biomass than FePISCES even though the iron concentration in FePISCES were greater than FeBFM. Furthermore, unlike the other HNLC environments, the EP was dominated by picophytoplankton in both configurations but FePISCES observed a greater biomass of diatoms than FeBFM. Though not much visible in the figure, the community compositions for the NAG were identical for both configurations with flagellates dominating the system. The NEP also saw both iron configurations attaining a similar total biomass and very similar community compositions. The only marginal difference was that FePISCES had a greater biomass of diatoms than FeBFM. Consequently, the elevated proliferation of diatoms in the HNLC environments could be attributed to the elevated iron concentrations. However, in order to understand the phytoplankton dynamics in more detail, it was necessary to analyse the time-series for the various species in order to understand how the community composition changed as the model simulations evolved.

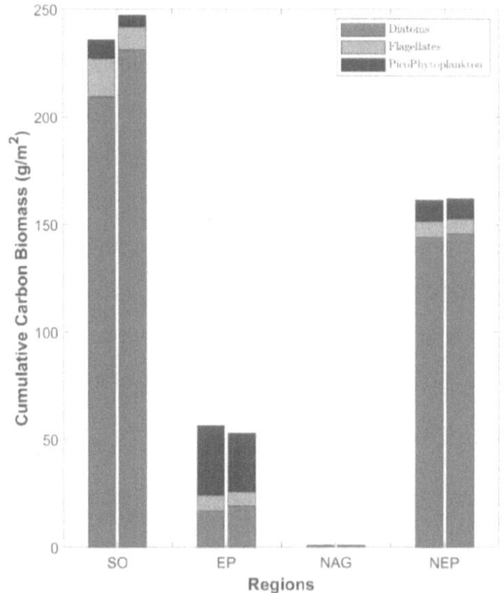

Figure 12: Cumulative carbon biomass for diatoms, flagellates and picophytoplankton after 10 years with FeBFM (left bars) and FePISCES (right bars)

Fig. (13) illustrates the differences in the phytoplankton groups between FeBFM and FePISCES in the modelled regions and complements Fig. (12) by illustrating the time evolution of biomass. Beginning in the SO, FeBFM reached an equilibrium state whereby the various phytoplankton groups maintained a similar seasonal cycle as the simulation progressed. While FePISCES saw a continued shift in the phytoplankton assemblage as diatoms grew in concentration which reduced the abundance of flagellates and picophytoplankton. The growing biomass of diatoms incidentally explained the trend in the remineralisation rate for FePISCES (Fig. 11), whereby a greater abundance of diatoms resulted in the elevated production of particulate organic matter as the model run progressed. To explain the overall elevated biomass in FeBFM compared to FePISCES in the EP (Fig.

12) it is important to note that, diatoms are the only phytoplankton group which require silicate as a macronutrient. In contrast to the SO and NEP (see Fig. 14 and 17), the EP had significantly lower concentrations of silicate (Fig. 15). Consequently, the elevated iron concentration in FePISCES spurred on a greater biomass of diatoms in comparison to the other phytoplankton groups. Whereas in FeBFM, the lower concentration of iron did not favour the proliferation of diatoms and instead favoured picophytoplankton. As picophytoplankton were not limited by silicate availability, this allowed FeBFM to achieve a greater biomass than FePISCES.

For the NAG, FeBFM and FePISCES had identical time-series for all the phytoplankton groups even though there was a stark difference in the ten-year iron time-series for the region (Fig. 10). This suggested that the behaviour of iron in the NAG for both configurations was controlled by similar biogeochemical processes (refer Sec. 4.2.3). Unlike the other HNLC regions, the NEP had a near identical phytoplankton biomass for FeBFM and FePISCES despite differences in dissolved iron (Tab. 3 and Fig. 10). Though the total biological productivity does not differ significantly between the two iron configurations, some regions observed changes in their dominance of a certain phytoplankton species. In addition, the non-linear response in biological productivity relative to the abundance of iron reflected the influence of additional nutrient limitations. Consequently, the next section evaluates the impact of the iron configurations on the cycling of macronutrients.

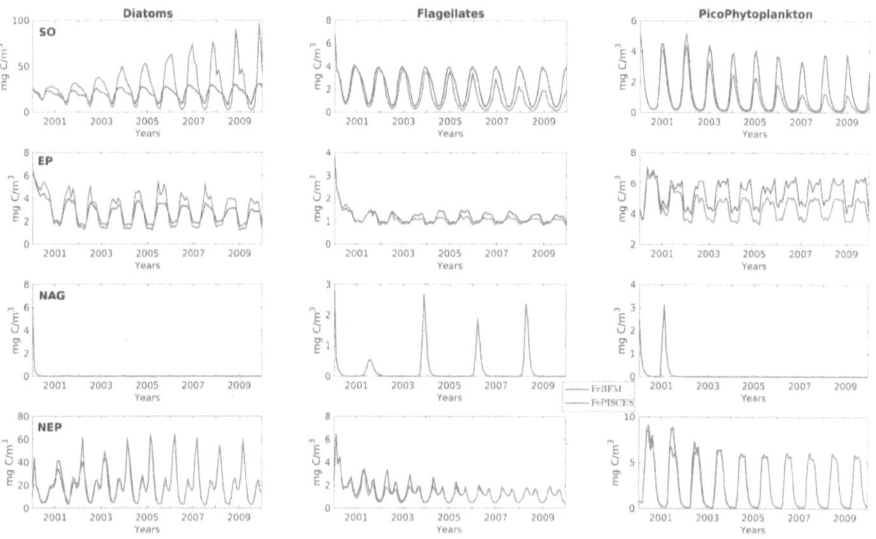

Figure 13: The time-series of the biomasses of diatoms, flagellates and picophytoplankton for FeBFM and FePISCES

4.2.2 Influence on the seasonality of macronutrients

As nutrients do not cycle independent of each other in a biogeochemical system it was important to understand whether changing the iron configuration would affect the seasonality of other major macronutrients such as nitrate, phosphate and silicate as well as chlorophyll. Therefore, each region would be analysed separately and the macronutrient as well as chlorophyll dynamics would be compared against observational data from the WOA18 and ESA. This was not intended for validation purposes but to understand whether changing the iron configuration would lead to a better correspondence with observational measurements.

Southern Ocean

Fig. (14) illustrated the typical HNLC conditions for the SO where there was an excess concentration of macronutrients accompanied by little chlorophyll due to limited iron availability (refer to Sec. 3.2.2). Both configurations had greater chlorophyll concentrations than the observed with FePISCES having a greater chlorophyll concentration than FeBFM due to the greater quantities of iron that were available (Tab. 3) which allowed diatoms to flourish (Fig. 13). Furthermore, both configurations showed a chlorophyll minimum in June/July, corresponding to the winter months and periods of little biological activity, but neither configuration accurately resolved the timing of the chlorophyll maximum. Because of the diatom growth in FePISCES: nitrate, phosphate and silicate were utilised for primary production which reduced their concentrations, but no nutrient was entirely depleted which permitted diatoms to continually grow throughout the model run (Fig. 13). This facet regarding nutrient depletion will become important in subsequent regions.

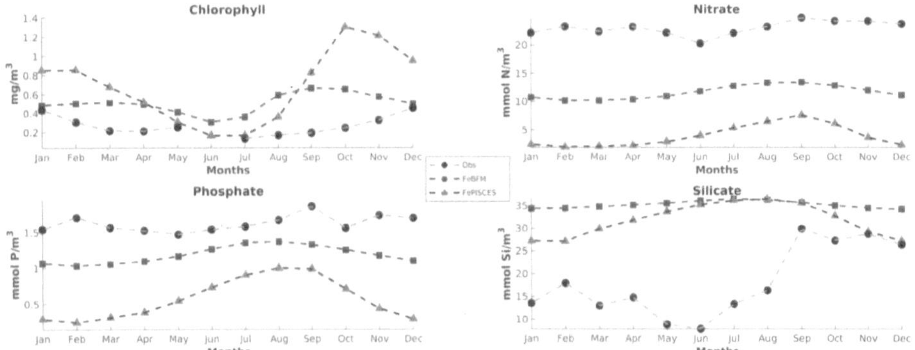

Figure 14: Monthly climatologies of: chlorophyll, nitrate, phosphate and silicate in the SO for FeBFM (blue) and FePISCES (red). The iron configurations were compared against observational data (black) from the WOA18 as well as the ESA

Furthermore, FePISCES saw greater seasonal variations in the macronutrient cycles than FeBFM. Though neither configuration captured the observed nutrient concentrations, FeBFM had more similar seasonal cycles for phosphate and nitrate relative to the observations. This was in major part due to FeBFM maintaining a low iron concentration (Fig. 10) consistent with observational measurements (Martin *et al.* 1991; Schlitzer et al. 2018).

Equatorial Pacific

Fig. (15) highlights the HNLC conditions prevalent in the EP, although the nutrient concentrations are lower than in the SO. The EP is a region heavily influenced by episodic upwelling events which are driven by the easterly trade winds (Sec. 3.2.1). As the BFM was run in its 0D, uncoupled state, this physical dynamic could not be captured. As a result of being unable to resolve the upwelling dynamics, the seasonality for chlorophyll for both the configurations were completely out of phase with the observations. In addition, neither of the iron configuration could resolve the observed seasonal cycle of nutrients, but they both developed a seasonal cycle of primary producers despite the low nutrient standing stock (Fig. 13). FeBFM maintained a slightly greater chlorophyll concentration which was reflected in the greater total biomass in Fig. (12). Referring back to Sec. (3.2.1), the premise of the simulations was to assume all the biological processes were constrained within the MLD. Consequently, the complete consumption of nitrate did not terminate biological productivity (Fig. 13) due to the system being sustained through regenerated production of ammonium. Therefore, though the EP attained almost double the iron concentration in the FePISCES configuration than in FeBFM, the limited availability of preformed nitrate hindered the biological productivity of both configurations.

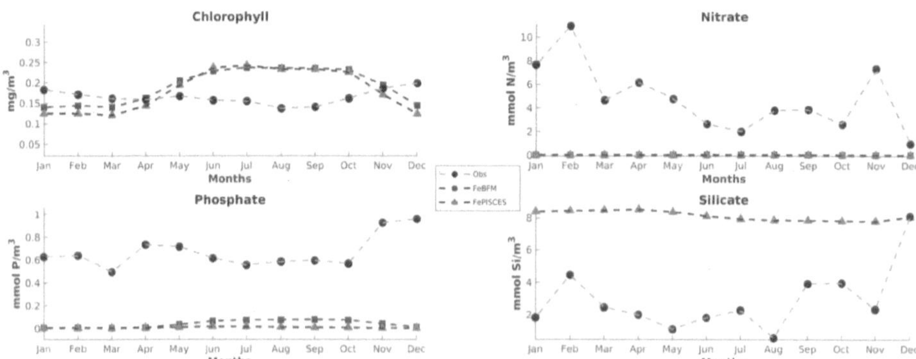

Figure 15: Monthly climatologies of: chlorophyll, nitrate, phosphate and silicate in the EP for FeBFM and FePISCES.

North Atlantic Gyre

The NAG is an oligotrophic region and has small nutrient concentrations and little biological activity at the surface as seen in Fig. (16). Both iron configuration showed nitrate and phosphate to be depleted for the entire duration of the simulation. As a consequence of the low macronutrient concentrations, the biological productivity was very low in both configurations (Fig. 13). Though the NAG is phosphate limited (refer to Sec. 3.2.1) which reduces biological productivity, the lack of seasonality in either the nitrate or phosphate concentration for both configurations suggests that the model was not directly capturing the nutrient dynamics for the region. Instead, the oligotrophic nature was observed due to the model being unable to simulate productivity at low nutrient levels. The NAG does illustrate the phenomenon that additional nutrient stresses affect the biological productivity of the region independent of the iron concentration.

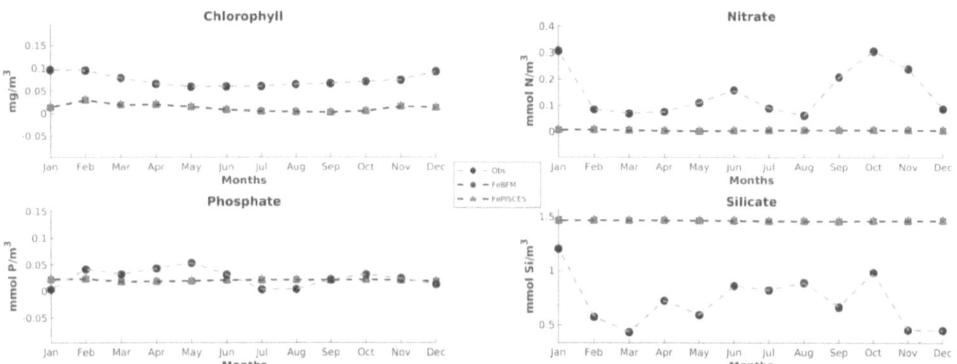

Figure 16: Monthly climatologies of: chlorophyll, nitrate, phosphate and silicate in the NAG for FeBFM and FePISCES.

North-east Pacific

The NEP was an interesting HNLC region (Fig. 17) as it had similar nutrient concentration to the SO but had a moderate flux of atmospheric dust. As expressed in Sec. (4.1.1), it was odd that the peak period of biological activity occurred in February/March, corresponding to winter, while in the observations this occurs in the summer months (Anderson 1969). According to the prescribed climatological boundary condition, the NEP region is characterised by a sharp increase in iron deposition from February (Fig. 8) which corresponds to the period of increased biological activity in the model. Consequently, the early onset of the chlorophyll maximum could have been due to the winter deposition of atmospheric dust that spurred on biological activity. Similar to the SO, both configurations struggled to capture the trend of chlorophyll. However, the NEP is a region which has a subsurface chlorophyll maximum that occurs at 55-65 m (Anderson 1969). Though within the range of the modelled MLD, satellite products would struggle to capture this feature. Again, the similar phytoplankton dynamics for both configurations could be attributed to the depletion of nitrate but the slightly higher concentration in diatoms in FePISCES could be seen in the lower concentration of silicate

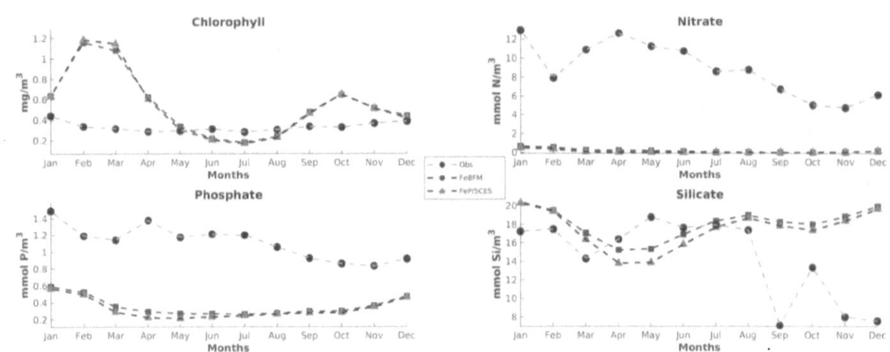

Figure 17: Monthly climatologies of: chlorophyll, nitrate, phosphate and silicate in the NEP for FeBFM and FePISCES.

4.2.3 Principal component analysis on key state variable relationships

In summarising the findings thus far, the various scavenging and remineralisation schemes employed in FeBFM and FePISCES did have a definite impact on the cycling dynamics of iron in the modelled regions. A major feature that distinguished the two configurations was the greater concentration of iron in FePISCES than FeBFM. However, besides for the SO, the elevated iron concentration had little impact on affecting the overall biological productivity in the remaining regions. This was due to the influence of additional nutrient limitations which resulted in similar biomasses. Consequently, as the phytoplankton assemblages were similar between FeBFM and FePISCES, this resulted in similar macronutrient cycles which corresponded poorly with observational data. Incidentally, the use of time-series plots was a useful technique in understanding how state variables evolved in the model simulations. But it was difficult to understand the relationship of one state when contextualized in the whole model system. Consequently, PCA (see Sec. 3.2.2) was chosen as a useful tool for identifying changes in key state variable relationships in a multi-variable system. When reading the PCA plots, refer to the list of in Tab. (4) and note that the scores represent monthly mean states. Furthermore, the DOC ($R_c^{(1)}$) variable was used as a proxy in FePISCES for the influence of ligands.

Table 4: List of symbols used in the PCA plots

Symbol	Description
Fe	Dissolved iron
POFe	Particulate organic iron
DOFe	Dissolved organic iron
DOC	Dissolved organic carbon
Chla	Chlorophyll
O2	Oxygen
N	Nitrate
P	Phosphate
Si	Silicate

Southern Ocean

Using two principal components (PCs), 78% of the total model variance was explained in FeBFM while 87% was explained in FePISCES (Fig. 18), with both configurations being strongly influenced by the first PC. In comparing the variable relationships between the two configurations, there was not a significant shift between FeBFM and FePISCES. Both configurations showed: nitrate, phosphate and silicate to be strongly correlated with each other; however, they were poorly correlated to iron. In addition, both configurations highlighted the poor correlation between bioavailable iron and its particulate organic species, however, it was strongly negatively correlated with its dissolved organic component. There was no shift in the relationship between DOC and iron from FeBFM to FePISCES. Treating the DOC content as a proxy for ligand concentrations, for FePISCES, ligands were not playing a major role in controlling the cycling of iron as there was no correlation between iron and DOC; however, DOC was strongly related to PC1 in FePISCES. Of significance was the fact that in FeBFM, iron explained a greater proportion of the model variance than in FePISCES where it was far removed from the other state variables. Focussing on the scores, FePISCES observed more clustering than in FeBFM.

Equatorial Pacific

Like the SO, the EP (Fig. 19) saw the macronutrients maintaining a similar relationship in both configurations. Both FeBFM and FePISCES were strongly influenced by PC1 with 91.5% of the total variance being explained by both PCs in FeBFM and 89.46% in FePISCES. Unlike the SO, both the organic iron species in FeBFM showed a positive correlation with iron which was odd considering that the production of particulate iron is accompanied by the consumption of bioavailable iron. However, DOC became more correlated with iron in FePISCES which suggested that organic ligands may have played a role in controlling the dynamics of the iron cycling. Instead, there is a decoupling between bioavailable iron and its dissolved organic species in FePISCES. Similar to the SO, FePISCES saw iron being far removed from the other state variables, suggesting that it did not play a significant part in influencing the function of the system.

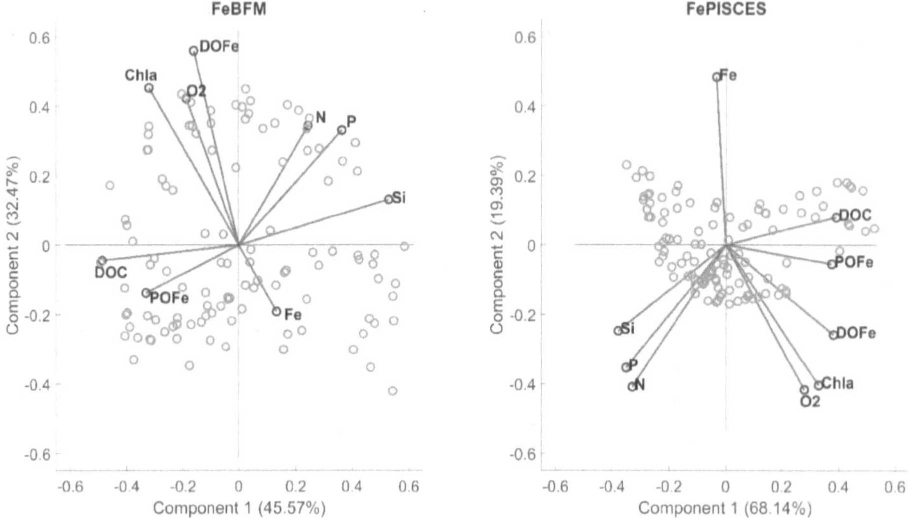

Figure 18: PCA plot for the SO with the variables indicated by a blue dot and the scores are shown with black dots with the orientation of each PC being arbitrary.

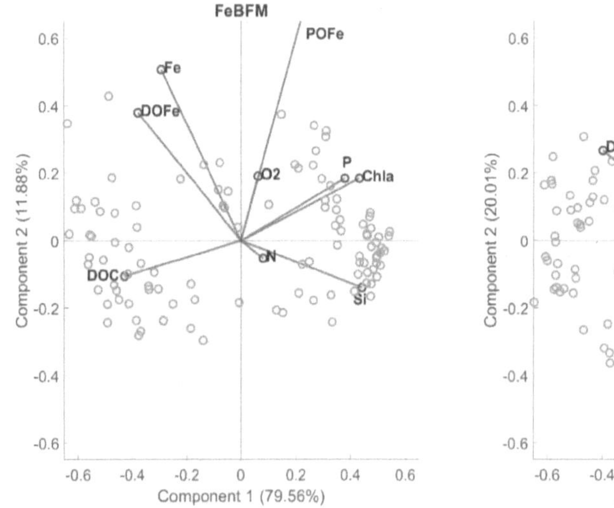

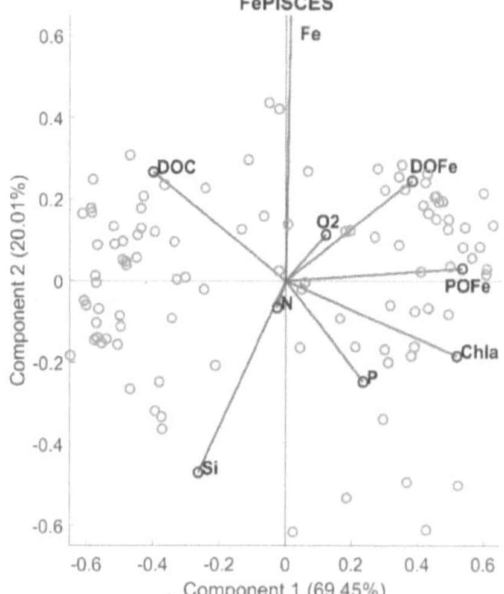

Figure 19: PCA plot for the EP

North Atlantic Gyre

For the NAG (Fig. 20), the PC1 and 2 explained approximately 78% of the variance in FeBFM and 82% in FePISCES, with both configurations being strongly influenced by PC1. In addition, the distribution pattern of the scores was very similar (the sign of PCs is arbitrary and it can be changed without affecting the decomposition) in both the configurations. From Fig. (20), bioavailable iron is not correlated with any of the other state variables in FeBFM and is not well explained by the first two PCs. However, in FePISCES, iron becomes strongly coupled to DOC and decoupled from both organic species as they are orthogonal to each other. This suggests that ligands were playing a role in affecting the cycling of iron; however, it was unclear whether ligands were responsible for the biennial cycle of iron seen in Fig. (10).

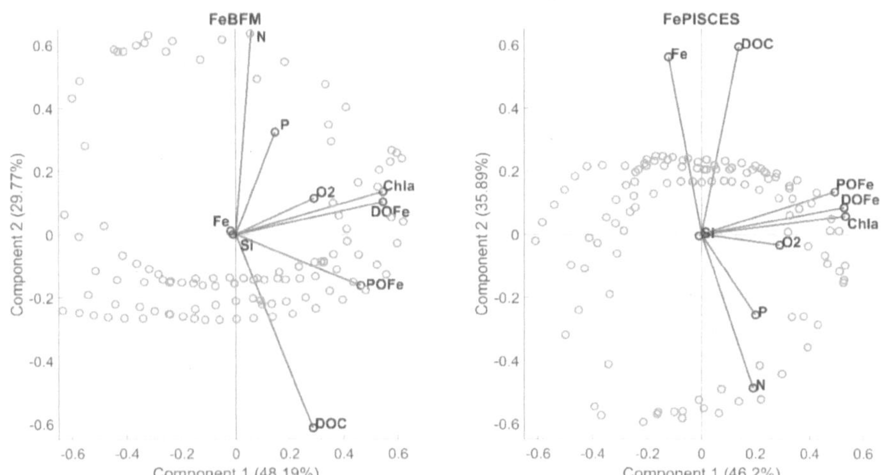

Figure 20: PCA plot for the NAG

North-east Pacific

The NEP (Fig. 21) was an interesting HNLC region because unlike the EP and SO, the NEP had a moderate dust deposition flux (Fig. 8) which resulted in a greater iron content for the region. In both FeBFM and FePISCES, the value of the respective PCs was similar which suggests that both configurations were constrained by the same biogeochemical processes. Both configurations saw a similar pattern for the distribution of the scores and there was no discernible difference between FePISCES and FeBFM in terms of the variable relationships. Focussing on iron, both configurations saw dissolved iron being negatively correlated with the particulate and dissolved organic species. However, the relationship between iron and DOC does not change from FeBFM to FePISCES. Noting the fact that the DOC content only affects the ligand concentration in FePISCES, it was unclear why the variable relationship remained the same.

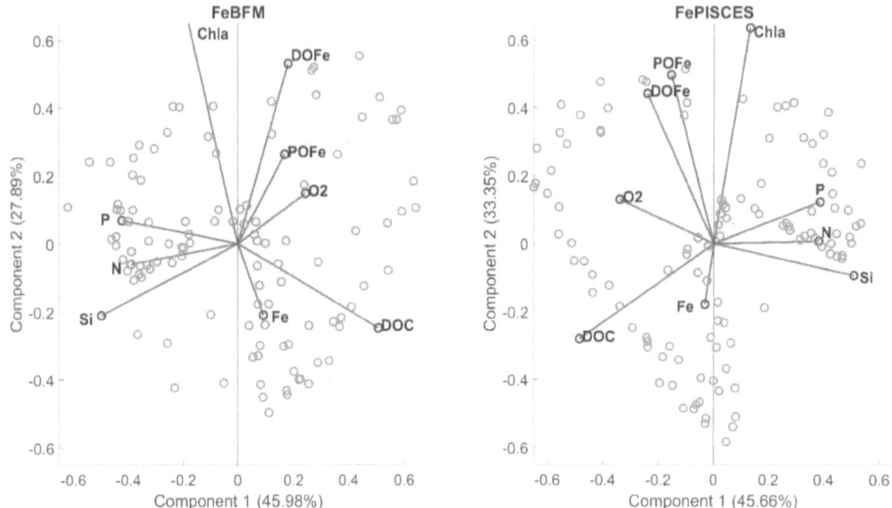

Figure 21: PCA plot for the NEP

4.2.4 Impact of ligands on the cycling of iron

The parameterisation of ligand dynamics were significantly different between the configurations (Sec. 3.1.3) and owing to their importance in affecting the scavenging regime, it was necessary to appreciate what affect the ligands were having on the cycling of iron. The PCA results hinted that the shift in the DOC variable to become more correlated with iron in FePISCES was due to the influence of organic ligands. However, the shift may have also been caused by the slight differences in the phytoplankton assemblages between the configurations which would result in different levels of DOC production (Eq. 2.10). Consequently, Fig. (22) shows the time-series evolution of DOC for each region and it will be used to further understand the implications of DOC on the cycling of iron. Fig. (23) will then be used to explore the relation between DOC on the scavenging regimes of both parameterisations. The section will end with an analysis of the scavenging parameterisation used in FePISCES in order to contextualise the implication of DOC on the free scavenging regime.

DOC time-series

Starting in the SO, it was anomalous that FeBFM maintained a small and stable DOC concentration while FePISCES maintained a similar DOC content to FeBFM until 2006. At this point the concentration spiked and continued to grow. Unlike FeBFM which established a steady-state regarding the concentration of diatoms (Fig. 13), FePISCES had continued growth. Therefore, the increase in the DOC content may have been due to the system accumulating DOC which is linked to the non-linear relationship of the PDEs. An explanation may be due to the system becoming nutrient stressed, implying that the greater iron abundance in FePISCES resulted in the overconsumption of key macronutrients, such as phosphate and nitrate, which resulted in the increased production of DOC due to the parameterisations of DOC in the BFM (refer to Eq. 2.10). Incidentally, the idea of nutrient stress was also seen in the EP, as the DOC content was greater than that of the SO. This suggested that most of the assimilated carbon existed in the dissolved pool rather than the particulate. Both configurations had similar phasing to each other with FePISCES maintaining a greater DOC content throughout the simulation than FeBFM. Consequently, the greater concentration of iron in FePISCES for the EP may have spurred on a greater diatom biomass resulting in a greater production of DOC.

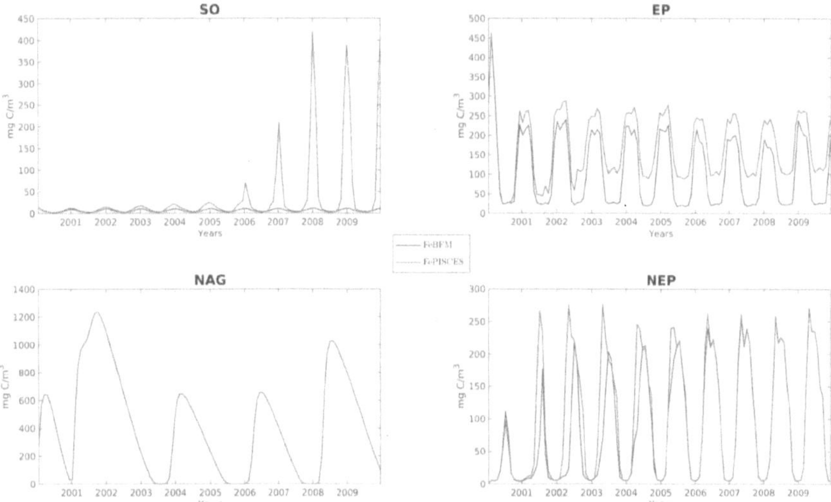

Figure 22: Time-series plot of DOC with FeBFM (red) and FePISCES (blue) represented for the four modelled regions

In Fig. (22) for the NAG, the DOC time-series for FeBFM cannot be seen as it is overlapped by FePISCES's. The DOC content of the NAG showed the same biennial oscillation for both FeBFM and FePISCES. Comparing the DOC time-series of FePISCES with the ten-year time-series of iron in Fig. (10) revealed a plausible relationship between the two variables as both exhibited a biennial oscillation. On closer inspection, the time-series of DOC is not perfectly in phase with that of iron, which suggests that there is not a linear relationship between the two variables, but this is understandable concerning the multiple non-linear relationships that affect both iron and DOC. Therefore, as both configurations had the same DOC time-series it provides some evidence that DOC may have been a major variable influencing the behaviour of iron in FePISCES. Consequently, the nutrient stressed environment of the NAG in the modelled regions resulted in almost all assimilated carbon being transferred to the dissolved pool (Eq. 2.10) which resulted in the largest concentrations of DOC for any region. The NEP had similar DOC concentrations to the EP with both FeBFM and FePISCES having a similar DOC time-series after the simulations adjusted (FeBFM was slower). The similar phytoplankton community structures between the two configurations (Fig. 12) would result in the near identical production of DOC (Eq. 2.10). Consequently, if DOC content was similar in FeBFM and FePISCES and as there was no shift in relationship between DOC and iron in the PCA analysis (Fig. 21), this does not suggest that ligands were not playing a role in affecting the iron cycling for FePISCES in the NEP. Instead, the moderate deposition of dust in the region would promote high iron concentrations (Tab. 3) and at the same time high biological productivity (Fig. 12 and 13) which could result in iron and DOC varying together regardless of the iron parameterisations.

Implications of DOC on the scavenging regimes

To further improve the understanding between the DOC dynamics and iron, Fig. (23) complements Fig. (22) by showing a correlation heatmap between DOC, iron and iron scavenging for the various modelled regions.

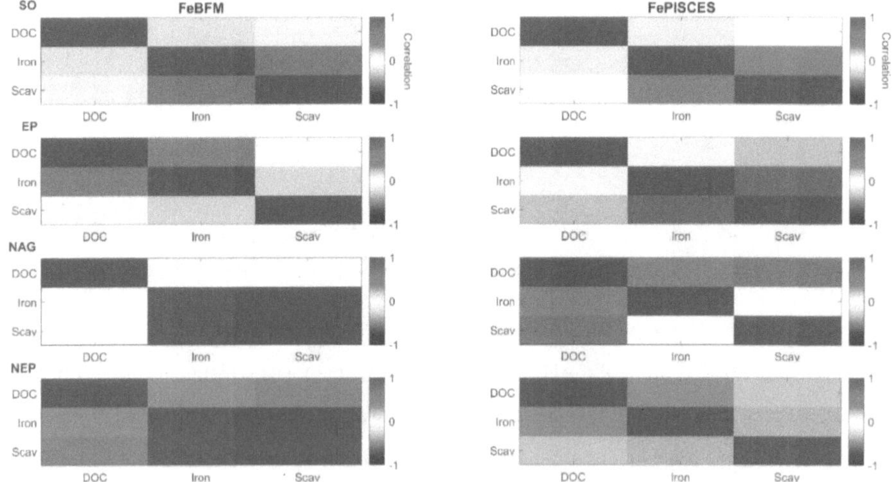

Figure 23: Correlation heatmap showing the relationships for: DOC, iron and scavenging (Scav) for FeBFM (left) and FePISCES (right) in the four modelled regions

In the SO, iron observed a small negative correlation with DOC and a strong positive relationship with scavenging in FeBFM. Furthermore, DOC was weakly correlated with scavenging which is understandable because the DOC dynamics do not implicate the scavenging regime in FeBFM. Contrasting FeBFM with FePISCES, there was a marginal shift in the correlation between DOC and iron as it became positive while DOC became negatively correlated with scavenging. Furthermore, there was a slight reduction in the correlation between scavenging and iron.

Like the SO, the EP also saw the correlation between DOC and scavenging becoming more negative from FeBFM to FePISCES but unlike the SO, the iron to DOC correlation decreased. However, the scavenging to DOC relationship became more negative from FeBFM to FePISCES. In the NAG, DOC was poorly related to iron and scavenging in FeBFM which was also seen in Fig. (20) for the PCA plots; while iron was strongly related to the scavenging rate. However, in FePISCES, DOC became significantly more related with iron but there was a major decoupling in the relationship between scavenging and iron. In addition, the scavenging to DOC relationship became more negative. Similar to the NAG, the NEP showed a strong positive relationship between DOC and iron for FeBFM as well as a positive relation to scavenging. In FePISCES, scavenging became negatively correlated with DOC and there was a reduction in the correlation between scavenging and iron.

From Fig. (23), a recurring phenomenon is the negative correlation observed between DOC and scavenging in FePISCES as well as a poorer relationship between iron and scavenging. Referring to Sec. (3.1.3), the concentration of ligands was calculated in relation to the DOC content using Eq. (2.17). Consequently, though the DOC content was similar in both configurations, the implications would be elevated ligand concentrations in the FePISCES configuration. Using Eq. (2.17), the mean ligand concentration over the ten-year period of the model simulations could be calculated for each region. The NAG had the highest mean concentration of ligands at 3.81 nM while the SO had the lowest (0.86 nM). The EP and NEP were similar with mean concentrations of 1.87 nM and 1.37 nM. Though the ligand concentration was merely proportional to the DOC content, the implications of a greater ligand concentration would influence the scavenging regime.

Starting with Eq. (2.16), $N^{(7)}$ can be made the subject of the formula of the quadratic by rearranging the equation to the following:

$$N^{(7)} = K^{eq}_{N^{(7)}}(N^{(7)}_{free})^2 + \Delta N^{(7)}_{free} \tag{2.21}$$

Noting that $N^{(7)}_{free} \ll N^{(7)}$, Eq. (2.21) could be simplified by removing the squared term.

$$N_{free}^{(7)} \approx \frac{N^{(7)}}{\Delta} \tag{2.22}$$

Expanding Δ in Eq. (2.22) gives:

$$N_{free}^{(7)} \approx \frac{N^{(7)}}{1+K_{N^{(7)}}^{eq}(L_T-N^{(7)})} \tag{2.23}$$

The simplified relation shown in Eq. (2.23) shows that the amount of iron available to scavenging is inversely proportional to the ligand concentration. Therefore, when the ligand concentration is high, the amount of iron that can be scavenged is small. But, when $L_T \approx N^{(7)}$ then the amount of iron that can be scavenged is directly proportional to the concentration of dissolved iron. Therefore, a greater ligand concentration should limit the ability of scavenging and inherently allow iron to accumulate as seen in Fig. (10). This idea is corroborated from Fig. (23) as the negative relationship between DOC and scavenging in FePISCES highlights that elevated ligand concentrations result in most bioavailable iron being complexed and thus unavailable to scavenging. While when the DOC content drops, the scavenging rate can elevate due to less bioavailable iron being complexed. Consequently, the DOC content for a region was indirectly driving the iron system by altering the scavenging regime of FePISCES.

5. Discussion

Multiple parameterisations and mathematical formalisms can be utilised to describe the iron cycle in biogeochemical models (Sec. 2.2.3) and therefore it is important to understand what implications different parametrisations have on the functional behaviour of a biogeochemical model in its ability to represent major processes such as: dissolved iron concentrations, macronutrient cycling and phytoplankton assemblages (Sec. 4). Therefore, this chapter is divided into two sections, with the first discussing whether the choice of iron parameterisations is significant when running a biogeochemical model and the second section exploring how useful 0D models are as spaces for testing different parameterisations.

5.1 Does the choice of iron parameterisations matter

In a biogeochemical model, the various PDEs used to describe key biogeochemical processes do not function in isolation. Instead they mathematically interact and influence each other which can result in non-linear behaviours and feedbacks. Consequently, by altering the iron parameterisations in BFM, it was expected that the functional behaviour of the BFM would change but to what degree could not be determined prior to performing the various experiments. It is important to note that when undertaking the book work, any free scavenging model variant from a different biogeochemical model could have been used. Therefore, this discussion does not constrain itself to solely comparing the inorganic iron dynamics of BFM and PISCES but also intends to be a general commentary on translating parameterisations into different biogeochemical models.

5.1.1 Inorganic iron parameterisations

From Sec. (4) the parameterisations of FePISCES produced dissimilar results to FeBFM when comparing the: concentration and seasonality of dissolved iron, scavenging and remineralisation rates, phytoplankton community compositions and macronutrient cycles in all the modelled regions. Contrasting the inorganic iron parameterisations of FeBFM and FePISCES, it was clear that the remineralisation schemes were more similar than the scavenging. Both configurations parameterised remineralisation as a linear process, which is the simplest form one can choose when data are not available to better constrain the parameterisations, with FePISCES including additional environmental stresses such as an oxygen dependency term which just modulated the intensity of the remineralisation scheme. The similarity in the remineralisation schemes was seen when analysing the rates in Fig. (11) as both configurations showed similar seasonal cycles for the remineralisation rates, with FePISCES having a higher rate than FeBFM which could be attributed to the parameter value used for the remineralisation rate constant.

In contrast, the scavenging regimes between FeBFM and FePISCES differed significantly and this was due to FePISCES employing a different model formalism for scavenging as well as including a prognostic appreciation of organic ligands which was linked to the concentration of DOC (refer to Sec. 3.13). The free scavenging model of FePISCES resulted in the low iron regions of the SO and EP having almost zero scavenging while in the high iron regions of the NAG and NEP, there was intermittent periods of intense scavenging followed by little to no scavenging (Fig. 11). The simpler model of FeBFM reached a steady-state in all the modelled regions and produced a clear seasonal cycle. The disparate scavenging regimes for the two configurations highlight the importance of constraining the scavenging rates which is a sentiment shared by other authors such as Tagliabue *et al.* (2016) and Yao *et al.* (2019). Especially in a free scavenging model like FePISCES, constraining the scavenging regime is necessary to avoid accumulation of iron, mainly in HNLC regions. A major reason for this is that in a 3D coupled model simulation, regions with excess iron concentrations would seed productivity in adjacent regions as iron would be advected and transported by the physical model. Incidentally, this could bias the modelled distribution of primary production since iron is a limiting nutrient in large regions of the global ocean.

A prognostic appreciation of organic ligands is a feature that is not common in current biogeochemical models (Tagliabue *et al.* 2016) and therefore it acted as an additional facet in the free scavenging model of FePISCES. Though the parameterisation of ligands was basic, the ligand concentrations in FePISCES were on average 1.4 nM greater than the prevailing iron concentrations in each region which reflects well with the observations of Gledhill & Buck (2012) in terms of representing the feature of uncomplexed ligands. However, coupling the ligand concentration to DOC resulted in the NAG having a greater ligand concentration than the SO. This result is not supported by observational evidence because the SO has greater biological activity than the NAG and owing to the production pathways of ligands (Hassler *et al.* 2017), the SO should have a greater concentration than the NAG. This highlights an issue with the DOC dynamics of the BFM. Therefore, the feedback in the model is one of mathematics which does not represent the biological behaviours of the modelled regions.

When adding new parameterisations to a model it is important to disseminate between mathematical feedbacks in the model system and process-based responses that would be expected in a real biogeochemical system. The PCA plots presented a useful tool in understanding how the variable relationships shifted with the choice of iron parameterisations. In general, for both iron configurations there was a decoupling of iron from the major macronutrients in the modelled regions. This could be due to the fact that iron is typically added as an additional multiple-nutrient limitation term within numerical models (Vichi *et al.* 2007b) and is not coupled to the functioning of other major nutrients. Not only was dissolved iron poorly related with the macronutrients but in most of the regions, it was weakly associated to its own organic species. In the SO, both iron configurations saw dissolved iron uncorrelated with its particulate organic species and negatively correlated with the dissolved organic species. Whereas in the EP, the improved relation of DOC to dissolved iron in FePISCES resulted in a decoupling of dissolved iron from its organic species. This same situation was seen in the NAG but not in the NEP as there was no shift in the relation of DOC and dissolved iron from FeBFM to FePISCES. Consequently, the ligand parameterisation of FePISCES caused a decoupling of dissolved iron to its organic species in some regions but both iron configurations struggled to resolve the biogeochemical relationships in the iron cycle.

As expressed in Eq. (2.23), the ligand concentration dictated the amount of iron available for scavenging in FePISCES. Using PCA, there was a strong relationship between DOC in the high iron regions of the NAG and NEP and from the similar biennial time-series of DOC (Fig. 22) and iron (Fig. 10) in the NAG, it could be inferred that DOC was indirectly driving the iron system by altering the scavenging regimes. This observation was corroborated using Fig. (23) where DOC became negatively correlated with iron scavenging in FePISCES. Therefore, as the DOC content increased, the amount of iron available to scavenging decreased. Furthermore, except for the EP, there was a reduction in the relationship between iron and scavenging in FePISCES which could have been due to the additional influence of DOC on the iron system. However, it is important to note that the study did not use the full PISCES model, but only the inorganic cycling of iron. Consequently, the spurious trends in iron observed for the FePISCES configuration may not occur in PISCES as the biology has been tuned appropriately and the production pathway of DOC may be suited to the free scavenging model. Furthermore, the formulation of iron dynamics in the full PISCES model of Aumont *et al.* (2015) contains full colloidal interactions and these parameterisations were not included in the test experiments of FePISCES. Incidentally, the accumulation of iron in the HNLC regions in the FePISCES configurations may have been due to the absence of a colloidal component.

5.1.2 Challenges of translating parameterisations into biogeochemical models

In addressing the question whether the choice of iron parametrisations is significant, the answer is yes. Using a different set of iron parameterisations will alter the biogeochemical behaviour of a model. Indeed, a major shortcoming of the book was that no parameter optimisation was done when translating the inorganic iron equations of PISCES into BFM so how much of the variance in dissolved iron concentrations could be attributed to the various model constants vs. model structure was unknown. This highlights a major challenge in translating parametrisations as not only the model formalisms need to be reconfigured to be suitable within a different model but also the various parameters may need to be appropriately tuned. However, appropriately tuning parameters will be difficult if there is not the necessary data to contain them which is a prevalent issue for iron parameterisations. In addition, if a parameterisation is sensitive to another state variable, such as ligands to DOC concentration, those facets may not be easily translated into a different model which means that the parameterisation may be a source error in model outputs. This further exemplifies the inflexibility in translating iron parameterisations between models.

When embedding new parametrisations, it is important to appreciate that a different model formalism of a biological process may not lead to improved model results. This does not disparage the fact that a more detailed parameterisation includes more meaningful biological interactions, instead, if the equations are not contextualised properly within a model, then the model performance could be worse off. Therefore, when translating parameterisations it would be advantageous to use parameter optimisation techniques to constrain parameters to optimal values (Ward *et al.* 2010) if sufficient data are available, but not the case for iron as well as conducting sensitivity tests to appreciate the various non-linear behaviours and responses. Ideally, the inclusion of more sophisticated parameterisations should be done only if they will lead to improved representations of a biological system, which become prevalent in coupled 3D simulations used for hypothesis testing.

5.2 Suitability of 0D models as spaces for testing parameterisations

The ultimate implementation of iron, which has a global valence, should be within coupled 3D models. However, large uncertainties are prevalent in describing the iron cycle in biogeochemical models (Yao *et al.* 2019). Furthermore, a wide variety of different parameterisations for processes such as scavenging and ligand dynamics are employed across the modelling community (Tagliabue *et al.* 2016). Acknowledging the work of Ménesguen *et al.* (2007) and McKiver *et al.* (2015), the choice of spatial resolution is not trivial when conducting biogeochemical modelling. Therefore, 0D models are dichotomous as their simplicity aids in the identification of non-linear processes and functional responses of parameterisations. However, a caveat to the simplicity of 0D models is the inability to include important physical processes which influence biology (McKiver *et al.* 2015). Therefore, this section wishes to explore the advantages and disadvantages of using 0D models for testing biogeochemical parameterisations by using the results of iron as a case study.

5.2.1 Advantages and disadvantages to 0D models for testing parameterisations

Running the BFM in its 0D configuration meant that important physical dynamics which implicate biological activity could not be included; however, this was to be expected given the limitations of a box-model set-up. For each region, it was assumed that the main source of exogenous iron was from atmospheric deposition and that the respective biogeochemical systems were sustained through regenerated production of nutrients, including iron, in the upper ocean, which was bounded between the surface and MLD.

Though some of the physical dynamics for the modelled regions could be accounted for with rudimental physical forcing functions and biological assumptions, the lack of a coupled physical model severely influenced the seasonal cycle of macronutrients, especially in the EP and NEP. In the EP, the chlorophyll and phosphate concentrations for the two iron configurations were out of phase with the WOA18 observations (Fig. 15) due to the inability to capture the upwelling dynamics of the regions. While for the NEP, the onset of a winter bloom in chlorophyll was attributed to the basic dust forcing function used in the model (Fig. 17). In addition, the NEP is a region with a sub-chlorophyll maximum (Anderson 1969) and this feature could not be represented in a mixed layer model. Furthermore, in all the modelled regions, nitrate and phosphate concentrations were drawn down below observed levels; however, in the 3D simulations of Vichi *et al.* (2007a, b), this same phenomenon was not prevalent. Low macronutrient concentrations will feedback onto iron as they will enhance N and P limitation in a system. Consequently, a weakness of 0D simulations seems to be the over utilisation of key macronutrients and this issue could be addressed by initialising the 0D model with greater macronutrient concentrations so that the steady-state resembles observational measurements.

0D models can be useful for testing and illustrating fundamental principles and ideas, with a good example being Daisyworld (Watson & Lovelock 1983). Though the 0D configuration of the BFM limited its ability to represent important physical dynamics, for the various modelled regions, a lot of iron features could be captured. This was accomplished through the addition of: a boundary flux for atmospheric dust, a particle tracer for dust and a variable MLD. However, the addition of multiple new components at once did not permit a step-wise assessment of the respective influence of each component on the modelling system. Consequently, a step-wise approach would involve multiple model runs whereby the various iron parameterisations are assessed while the number of additional modelling components such as a dust flux or variable MLD are incrementally added to assess the impact of the added modelling components on the iron system and to better disseminate the responses of the iron parameterisations.

Incidentally, as numerical models become more complicated by increasing the number of: dimensions or coupled components (physical, climate and ice models); the modelling system becomes more complex and thus more resilient to non-linear behaviours in parameterisations. Consequently, the complexity of a 3D model would make it difficult to test such sensitive parameterisations as iron and the different responses of the biogeochemical parameterisations may be diluted by the interactions with the physical model. In comparison, 0D simulations have a simpler structure which helps to better understand the sensitivity to different parameterisations. As seen with the identification of the variable relationships between FeBFM and FePISCES, it is unclear whether a 3D model would be able to identify the non-linear responses in the iron system, but within a 0D context, the identification of non-linear behaviours was aided due to the simpler model structure. At present, the only method to improve both 0D and 3D representations of iron and its impact on the wider biogeochemical systems of the ocean lies in improved observational data. At present, the GEOTRACES program has successfully increased the number of iron observations but what is lacking are seasonal measurements of dissolved iron. Having seasonal observations of iron would allow an assessment of how iron evolves with other key macronutrients. From increased observational data, existing iron parameterisations could be improved, with 0D models being an ideal space to test and refine them before being implemented in coupled 3D simulations.

6. Conclusion

6.1 Conclusion

This book has compared and contrasted the functional behaviour of iron parameterisations implemented in current biogeochemical models. The iron cycle has only recently been included in biogeochemical models and the parameterisations that describe key processes such as scavenging and ligand dynamics are not well constrained by available observations. Therefore, there is a need to understand the biogeochemical modelling implications of using different iron parameterisations while new data are being collected. Using the BFM as the background numerical model the iron parameterisations of BFM and PISCES were tested in four regions: Southern Ocean, Equatorial Pacific, North Atlantic gyre and North-east Pacific; encompassing a diverse array of biogeochemical environments. From Sec. (4), the disparate functioning of FeBFM and FePISCES was evident as FePISCES observed significantly greater iron concentrations than FeBFM in all the modelled regions. This had implications on the phytoplankton assemblages as well as the macronutrient cycles of FePISCES in all the modelled regions. The main difference in the iron formalisms of FeBFM and FePISCES was the scavenging regime and thus the scavenging model of FePISCES was implicated in the dissimilar iron concentrations. Furthermore, the ligand parameterisations of FePISCES lead to an apparent decoupling of dissolved iron from its organic species where in fact the inclusion of dynamical ligands should have led to the opposite.

Though the book did not aim to validate the various parameterisations, it is important to note that neither of the iron parameterisations was able to represent the key variable relationships involved in the iron cycle, namely the relationship of dissolved inorganic iron to its organic species. This represents a shortcoming in the application of these parameterisations in higher order models. In addition, the diverse behaviour of the different iron parameterisations potentially showcases a lack of consensus in the modelling community in the representation of dissolved iron because if the parameterisations are drawn from the same scientific knowledge, it would be expected that they would behave in a similar or near similar manner. Another facet to the study was acknowledging the difficulty in testing and translating iron parameterisations from one model to another, noting that various parameter values may need to be tuned appropriately as well as the sensitivity of parameterisations to other biogeochemical processes which may not become apparent until sensitivity tests are conducted. Therefore, the testing of parameterisations should be done within 0D models in order to assess any non-linear behaviours and ultimately embedded in 3D models to study how they interact with physics. Therefore, the choice of iron parameterisations in a biogeochemical model is significant as they will have different implications on model outputs.

6.2 Future work

In the book, no parameter optimisation techniques were applied when translating the PISCES iron parameterisations into BFM. Incidentally, how much of the variation in the outputs could be attributed to inappropriately tuned model parameters vs. model structure was unknown. Consequently, future studies could use the work of Ward *et al.* (2010) or Annan *et al.* (2005) in applying parameter optimisation schemes to numerical models. If parameter optimisation was done on the translated parameterisations, then with greater certainty, variations in model outputs could be attributed to the differences in model formalisms which would aid in the improvement of parameterisation schemes.

Another interesting metric that could be applied in testing different parameterisations is the computation of Lyapunov exponents. Lyapunov exponents are used in the study of non-linear systems (Wolf *et al.* 1985; Das 2012) to quantify how model trajectories diverge in phase space whose initial states are slightly different. Therefore, perturbing a biogeochemical model with a different set of iron parameterisations and understanding the time evolution and divergence of the whole model system would allow for a greater appreciation of the impact of different iron parameterisations on a biogeochemical model.

www.ingramcontent.com/pod-product-compliance
Lightning Source LLC
LaVergne TN
LVHW040909150826
845672LV00007B/1967